FROM BUSH-BURN TO BUTTER

• A JOURNEY IN WORDS AND PICTURES •

• BY ERIC WARR •

Butterworths

Wellington
1988

New Zealand	Butterworths of New Zealand Ltd, 205-207 Victoria St, CPO Box 472, WELLINGTON and NML Plaza 41 Shortland St, PO Box 2399, AUCKLAND
Australia	Butterworths Pty Ltd, SYDNEY, MELBOURNE, BRISBANE, ADELAIDE, PERTH, CANBERRA AND HOBART
Canada	Butterworths Canada Ltd, TORONTO and VANCOUVER
Ireland	Butterworths (Ireland) Ltd, DUBLIN
Malaysia	Malayan Law Journal Sdn Bhd, KUALA LUMPUR
Singapore	Butterworth & Co (Asia) Pte Ltd, SINGAPORE
United Kingdom	Butterworth & Co (Publishers) Ltd, LONDON and EDINBURGH
USA	Butterworth Legal Publishers, ST. PAUL, Minnesota, SEATTLE, Washington, BOSTON, Massachusetts, AUSTIN, Texas and D & S Publishers, CLEARWATER, Florida.

National Library of New Zealand
Cataloguing-in-Publication data

Warr, Eric.
    From bush burn to butter: a journey in words and
pictures / by Eric Warr. Wellington [N.Z.] : Butterworths,
1988.
    1 v.
    ISBN 0409787736
    1. Dairying—New Zealand—History. I. Title.
    637.09931

ISBN 0 409 787736

© **Butterworths of New Zealand 1988**

Printed in Singapore by Singapore National Printers Ltd

CONTENTS

ACKNOWLEDGEMENTS

The Publishers, on behalf of the author, wish to gratefully acknowledge permission to reproduce photographs from the following institutions:

Alexander Turnbull Library pages 3, 4, 5, 6, 12, 15, 17, 21, 25, 39, 43, 44, 47, 48, 50, 53, 54, 55, 56, 59, 63, 64, 66, 74, 80, 81, 82, 83, 93, 99, 100, 103, 104, 107, 112, 115, 117, 122, 123, 127, 128, 133, 134, 141, 144, and cover.

Auckland Institute and Museum pages 35, 37, 38, 40, 60, 70, 75, 76, 98, 99, 108, 116, 118, 128, 131, 163 (all of which originally appeared in the Weekly News).

Canterbury Museum pages 25, 26, 27, 28.

Canterbury University Library pages 28, 37, 119.

Dannevirke Public Library pages 11, 49.

Early Settlers Museum pages 31, 66, 72, 126.

Glaxo Laboratories page 160.

Hocken Library pages 23, 30, 32, 56, 61, 94, 111, 118.

Kiwi Co-operative Dairies pages 147, 165.

Massey University 46.

Morrinsville Co-operative Dairy Co page 46.

National Archive pages 1, 2, 8, 18, 65.

National Art Gallery pages 9, 34

Nelson Provincial Museum pages 22, 72.

New Zealand Co-operative Dairy pages 91, 145, 148, 149, 150, 151, 152, 153, 159, 161.

New Zealand Dairy Board pages 78, 157, 158.

New Zealand Forest Service page 45.

Palmerston North Public Library pages 10, 14, 139.

Taranaki Museum pages 41, 68, 73, 87, 89, 90, 121.

Waikato Museum of Art and History pages 69, 85, 100, 101, 120, 130.

Special thanks are also gratefully extended to the staff of the Faculty of Geography at Massey University, most particularly to Mrs June Brougham and to Professor Keith Thomson; and to the New Zealand Founders Society.

PUBLISHER'S NOTE

The late Eric Warr was Senior Lecturer in Geography at Massey University from 1959 until his untimely death in February 1988.

One of his most passionate interests was the history and development of the New Zealand dairy industry. He spent many years courageously battling ill health, researching the subject and gathering photographic material for the book he intended to write. *From Bush-Burn to Butter* is the result of his years of research and labour. His death in early 1988 meant that Eric could not be involved in the latter stages of the book and some information has unfortunately been lost to us, most notably his bibliography. The resultant work is, however, a fitting memorial to a man who worked so hard to uncover the industrial developments and social history of an industry which today is still one of New Zealand's most important income earners.

CHAPTER

1

Missionaries

AND MISSION STATIONS

The early missionaries and mission stations made a significant contribution to the development of farming in New Zealand. In 1814 on instructions from Marsden, some of the first dairy cows were imported for Church of England mission stations in the Bay of Islands. From the outset Marsden envisaged that the missionaries would not only introduce their Christian teaching but also create a greater measure of self-sufficiency in food supply by developing agriculture on the mission stations. Furthermore, he stressed the need to introduce to the local Maoris the improved practices and methods of European style farming. The same policy was adopted by the Church Missionary Society and by the Wesleyan missionaries.

Samuel Marsden encouraged mission station farming and was responsible for the first importation of dairy cows into New Zealand.

North Auckland Mission Stations

The establishment of all mission stations in the North Auckland area followed the same pattern — after a dwelling had been built the next priority was to clear, break up and cultivate a small area of land in grain and vegetables. Marsden encouraged the missionaries to extend their farming activities and at times chided them for insufficient efforts in this direction. He not only advised and admonished but also forwarded to them good quality livestock and seed from his Australian farms.

By 1820 a general farming pattern had emerged. Grain, especially wheat, was the main crop; within 18 months of the initial settlement of Rangihoua wheat was growing. Although there is little statistical data of early cropping, evidence suggests there was never much more than 10 acres planted on any one mission station. In the mission station gardens a variety of English vegetables and fruits were grown: potatoes, peas, beans, carrots, turnips, onions, cabbages, cucumbers, asparagus, rhubarb, melons, oranges, lemons, peaches, apples, cherries, plums, pears, quinces, Cape gooseberries, walnuts, figs, currants, raspberries, strawberries and grapes.

Cattle were the most important livestock on early mission farms. The cattle had originated from Australia and had been sent by Marsden; after 1814 he sent further stock from New South Wales at his own expense as the opportunity arose. In 1823 Marsden finally presented the herd to the Church Missionary Society. In 1821 25 head were recorded, increasing to 95 by 1829. In addition there were others running free but too wild to capture. The cattle wandered freely, grazing in the scrub and bush and correspondence suggested that 50 to 100 acres were needed to support

1

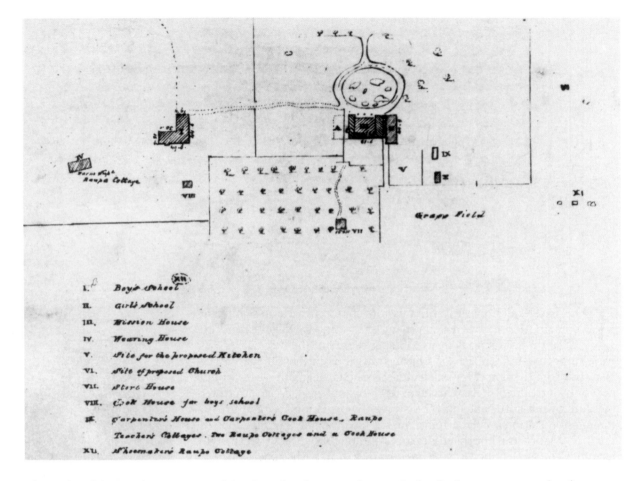

I.   *Boys School*
II.   *Girls School*
III. *Mission House*
IV.   *Weaving House*
V.   *Site for the proposed Kitchen*
VI.   *Site of proposed Church*
VII. *Store House*
VIII. *Cook House for boys school*
IX.   *Carpenters House and Carpenters Cook House, Raupo*
      Teachers Cottages. Two Raupo Cottages and a Cook House
XII. *Shoemakers Raupo Cottage*

Above: Plan of the Otawhao mission station.

Opposite page above: "Te Kohu Kohu", home of early settlers on the Hokianga River. Watercolour from the journal of Lt Morton Jones.

Opposite page below: The Otawhao mission house, school and mission station.

each beast under these conditions. Individual missionaries often kept one or two cows to supply their families with milk.

The numbers and kinds of particular livestock were largely determined, and limited, by the availability of feed or grazing. Thus while the sites of most early mission stations lay in scrub or fernland, there was little difficulty in keeping cattle or raising pigs and goats. Sheep created problems of their own.

Mission farming was, however, never completely successful. At no time in the 1820s were the mission stations wholly self-sufficient in food supply. Missionaries not only had notoriously large families to support but there were numbers of Maori "scholar-servants" as well, and few missionaries had experience in farming. The early mission sites were chosen for the protection and security they afforded the mission staff rather than for the potential farming quality of the area. Richard Davis, an exception in that he had farming experience, recognised this limitation soon after his arrival.

Waimate: New Zealand's Pioneer Farm

Before the establishment of the mission farm at Te Waimate in April 1831, the agricultural activities of early missionaries were fairly small-scale, providing only minimum subsistence. This was essentially "gardening" rather than farming — small cultivated patches of wheat and potatoes, gardens of English vegetables and fruit trees, with cattle and pigs running free and foraging in the nearby bush and scrub. Of the various missions the most persistent agricultural effort was made by the Church of England which set up fairly settled communities with large numbers to support.

English mission station at Kerikeri 1824. Taken from Capt LJ Dupperrey's "Voyage Autour du Monde".

Richard Davis, a successful farmer from Dorset, arrived in 1824 to take over and supervise mission farming activities. An earlier appointment to a similar position had proved notably unsuccessful. After early efforts at Kawakawa, Davis helped to establish the Waimate mission farm in 1831. On 250 acres purchased from the Maoris, he created a model English mixed farm, running livestock and growing crops. Davis took his farming seriously, for along with imports of farm implements and the like, he requested copies of up-to-date books on agriculture in order to keep up with all developments and improvements in farming. The initial success of this pioneer mission farm was due mainly to the energy, enterprise, and expertise of this missionary farmer.

Waimate's volcanic clay-loam soil made it an area of agricultural potential. Although originally covered in forest when taken over from the Maoris, the farm consisted of former Maori cultivated or crop land covered in tall bracken, fern and scrub with scattered patches of relict bush. The process of clearing the land proved long, arduous and costly. Fern and scrub were cut down with bean hooks, roots were grubbed up with hoes, and the larger stones were dug up and removed. Then the breaking-in plough was applied, but the land was still full of fern roots and stones which clogged up the plough, causing frequent delays. Under such conditions less than an acre was ploughed each day and further ploughing and clearing were needed before the land was ready for seed.

At Waimate, as on most earlier mission stations, wheat was the single most important crop. In 1831, there were 2.5 acres of wheat, increasing to 40 acres by the 1838-39 season. The actual area cultivated each year was severely limited by the inability of the mission to get Maori labour for land clearing, cultivating, sowing and harvesting. Fairly low yields were common. The wheat was hand ground in small, steel mills but after 1834 there was a flour mill, the first water-powered mill in New Zealand.

Dairy cows at Waimate at first gave little milk because there was a lack of suitable pasture grasses for grazing. Most of the Waimate livestock had originated on Marsden's farm in New South Wales and were of the highest quality. Such standards of quality in stock were, however, not maintained and later shipments from Australia to settlements in

Paihia CMS mission station and settlement 1827. Engraving by de Sainson, a member of d'Urville's crew.

Wellington, New Plymouth and Nelson were of much poorer quality. The expansion of dairying at Waimate was also limited by the lack of sown pastures but by 1839 100 acres were recorded, 80 acres of which were fenced-in paddocks with wooden post and rail fences. Such grasses were supplemented by lucerne and clovers.

The lack of suitable farm labour continued to prove a severe constraint upon farming development and was a factor in the ultimate decision to abandon the farm in the late 1830s, for it was proving too costly to run and maintain. In 1842 the farm was acquired by Bishop Selwyn to train young Maoris for holy orders. His disinterest in its farming function soon resulted in the farm falling to rack and ruin. During the war of 1845 it was occupied by British troops.

As the first pioneer and mission farm, the benefits and contributions of Waimate were limited. The local wheat and flour demand was never fully met. It occasioned little curiosity among the local Maoris, they preferred their traditional farming methods as opposed to those used on the model farm, although the building of the flour mill led to some interest in growing wheat. There was slight interest in cattle with only one or two head being kept by the Maori tribes.

On a visit to Waimate in 1835, Charles Darwin remarked on its "old world" appearance, "... as a fragment of Old England, small indeed but

Waimate mission station and farm — the first dairy herd.

genuine". The lack of lasting success was probably due to failure to appreciate the real differences in the New Zealand situation and consequently the continuing effort to apply wholly English, and frequently inappropriate approaches. By the time such differences were appreciated it was too late to act upon them effectively, because cheaper forms of farming had been developed to meet the special needs of the local markets and of settlers with very limited capital.

CHAPTER

2

EUROPEAN SETTLERS

Afeature of all early settlements around the coast of New Zealand was the presence of dairy cattle, not in large herds but in small groups of one or two sufficient to meet the dairy needs of individual colonial families. The ubiquitous cow was to be seen grazing freely wherever feed was readily available and was seldom, if ever, fenced in. While many of these cattle came from the large herds across the Tasman in the colony of New South Wales, some arrived on immigrant vessels as ship's cows, providing milk on board during the long voyage.

Agriculture in the Settlements: 1851 — North Island

In 1840-51 pioneer agriculture was dominated by two principal types of farming: semi-intensive mixed farming and extensive pastoralism. Cattle were associated with both. In the former bush districts of the North Island, and in the South Island scrub and fern areas, cattle and mixed farming were largely associated with dairying. By contrast, on the more open grasslands of the South Island and parts of the southern North Island, cattle on the tussock plains were run either on large cattle stations or grazed on runs in association with sheep. In the mixed farming areas, although there were some sown pastures available for grazing, cattle depended heavily upon foraging for feed.

Auckland

The population of Auckland in 1841 was approximately 1,200, rising by the end of 1842, to nearly 3,000 of whom about 1,800 lived in the town proper. By the end of 1842 the town was spreading beyond its original limits and settlement was extending into the "agricultural" suburbs in Epsom, Freemans Bay and Tamaki, while there were further scattered homesteads to the south at Papakura. Cattle and crops were raised to supply milk, meat and other foods. Some agricultural workers lived in the town proper, walking to work on nearby farms as there were no formal roads. In the early 1840s several rural lots were sold around the town and by 1842 about 300 acres were under cultivation, much of it land close to the track linking the Waitemata and Manukau Harbours. By 1851 the Auckland settlement had grown and expanded to cover the whole of the Tamaki isthmus, and the area within 8 miles of the township was intensively farmed, while less intensive settlements were scattered as far as Papakura.

NEW ZEALAND COMPANY—STEERAGE DIETARY.

FOR EACH PERSON FOURTEEN YEARS OLD AND UPWARDS.

	Prime India Beef.	Prime Mess Pork.	Preserved Meat.	Biscuit.	Flour.	Rice.	Preserved Potatoes.	Peas.	Oatmeal.	Raisins.	Suet.	Butter.	Sugar.	Tea.	Coffee.	Salt.	Pepper.	Mustard.	Vinegar or Pickles.	Water.
	lb.	lb.	lb.	lb.	lb.	lb.	lb.	pint												quarts
Sunday	—	—	$\frac{1}{2}$	$\frac{3}{4}$	$\frac{1}{4}$	—	$\frac{1}{4}$	—	One Pint Weekly.	Eight Ounces Weekly.	Four Ounces Weekly.	Eight Ounces Weekly.	Sixteen Ounces Weekly.	Two Ounces Weekly.	Two Ounces Weekly.	Two Ounces Weekly.	A Quarter-Ounce Weekly.	Half-an-Ounce Weekly.	Half-a-Pint Weekly.	3
Monday	—	$\frac{1}{2}$	—	$\frac{3}{4}$	$\frac{1}{4}$	—	—	$\frac{1}{4}$												3
Tuesday	$\frac{1}{2}$	—	—	$\frac{3}{4}$	$\frac{1}{4}$	$\frac{1}{4}$	—	—												3
Wednesday	—	—	$\frac{1}{2}$	$\frac{3}{4}$	$\frac{1}{4}$	—	$\frac{1}{4}$	—												3
Thursday	—	$\frac{1}{2}$	—	$\frac{3}{4}$	$\frac{1}{4}$	—	—	$\frac{1}{4}$												3
Friday	—	—	$\frac{1}{2}$	$\frac{3}{4}$	$\frac{1}{4}$	—	$\frac{1}{4}$	—												3
Saturday	$\frac{1}{2}$	—	—	$\frac{3}{4}$	$\frac{1}{4}$	$\frac{1}{4}$	—	—												3
Total Weekly	1 lb.	1 lb.	$1\frac{1}{2}$ lb.	$5\frac{1}{4}$ lbs.	$1\frac{3}{4}$ lbs.	$\frac{1}{2}$ lb.	$\frac{3}{4}$ lb.	$\frac{1}{2}$ pint	1 pint	8 oz.	4 oz.	8 oz.	16 oz.	2 oz.	2 oz.	2 oz.	$\frac{1}{4}$ oz.	$\frac{1}{2}$ oz.	$\frac{1}{2}$ pint	21 qrts.

Children Seven Years old and under Fourteen receive each, of Water, Three Pints a day; of other Articles, *Five-Eighths* of the Ration of an Adult.

Children One Year old and under Seven receive each of Water, Three Pints a day; of Preserved Milk, a Quarter-Pint a day; and of other Articles, *Three-Eighths* of an Adult's Ration; or, if directed by the Surgeon, either Four Ounces of Rice or Three Ounces of Sago, in lieu of Salt Meat, three times a week.

Infants under One Year old do not receive any Ration; but the Surgeon is empowered to direct an Allowance of Water, for their use, to be issued to their Mothers. No charge is made for their Passage.

The several Articles of Diet may be varied from time to time, under the direction of the Surgeon, so as to promote the health and comfort of the Passengers, especially of Children.

New Zealand House, 9 Broad Street Buildings,
London, 1st November, 1848.

The monotonous diet that steerage passengers could expect on the New Zealand Company ships.

The sown grass area in the Auckland settlement was of special significance, for in 1851 the settlement comprised nearly 60% of New Zealand's total sown grass area, with 70% of its total cultivated area in sown English grasses, the most popular being ryegrass.

The farming landscape surrounding the settlement was itself quite distinctive — solid stone walls, stone houses, and a variety of other stone structures. Hawthorn hedges were beginning to replace the earlier wooden post and rail fences. Local farms varied significantly in size, ranging from 100 to 1,000 acres, with field sizes varying equally as much, the largest occasionally as big as 200 acres.

Unlike other settlements during this period, cattle slightly outnumbered sheep. In 1851 there were 11,075 cattle and 10,943 sheep, possibly a reflection of the greater ease of cattle breeding within the Auckland settlement area.

Early Auckland Dairying: Ex-Regular and Militia Settlements

Auckland developed as a garrison town deriving much prosperity from the fact that troubles with the local Maori population tended to prejudice neighbouring farming developments. Militia settlers were a feature of early Auckland. In many North Island area's bushland was freely available for settlement and was virtually given away in grants in many insecure frontier districts. The first settlements of this type in the Auckland area were in 1847-48 when about 700 "fencibles" or British ex-regulars were settled south of Auckland in the Onehunga, Howick, Panmure and Otahuhu districts. These ex-regulars made up the local militia units ready

Compton's farm, Hutt Valley.
Sketch by William Swainson.

for action at short notice in any emergency. Each member was given an acre of land and settled in "pensioner villages" from which they were often also hired out to work on nearby farms.

An early effort at co-operation in dairying associated with one of these groups was the establishment of the Howick Pensioners Co-operative Cow Company on 29 May 1848. The company was founded by soldier settlers from England and consisted of 40 members each holding a £10 share. Its modest objective was to purchase one cow for each member. Membership of the company was open to anyone by paying an initial £1 the remainder being paid by instalments. The executive committee of 13 was committed to buying cows as funds were available, and selling the dairy produce to members at an agreed fixed price; surplus produce was made available for public sale at a fixed price. The cattle purchased were to be branded and registered with details being kept in a form of herd book. Disappointingly the co-operative failed when members neglected to keep up their weekly instalments.

Bush Farming in Taranaki — Bush-Burn Forest Clearance

In 1848, GB Earp described in detail the process of forest clearance by bush-burn. He estimated the overall cost of cutting, lopping and burning the bush to be about £3 an acre, but quickly added that ". . . timber in

Early settler's whare in the Manawatu, late 19th century.

locations of easy access will repay this sum". He went on to add that "... the stumps themselves it will be folly to clear away till they are so far rotted as to permit being dragged out with oxen". The clearing of the bush was soon recognised as a process demanding great skill, the actual burn being the trickiest phase of the whole operation.

Earp describes the first part of the bush clearing process as underscrubbing, or clearing away the supplejacks, lianas, small trees and ferns with a bill-hook. In the second part of the process, the smaller trees from 6 feet to 20 feet in circumference were felled with axe or cross-cut saw. This vegetation was left to dry and wither before the bigger trees were felled. Next Earp added, "... The pines should be thrown last because then they do not cover up the other stuff until it is dry. It is necessary to be very particular in lopping the branches of the trees when they are down, for if it does not lay all snug and compact you will not have a good burn. After it has been felled from one to three months you fire it in a strong wind.... Now comes the cleaning up; this is the worst job of all for everything is black and dirty. You chop up everything of reasonable size and collect it, together with what the fire has left, against the fallen monsters ... and burn it." Grass seed was then broadcast or sown by hand and the first crop was often prodigious giving excellent feed in a a very short time. "... If you are going to sow wheat or mangles or plant potatoes you take a little more trouble with the clearing up."

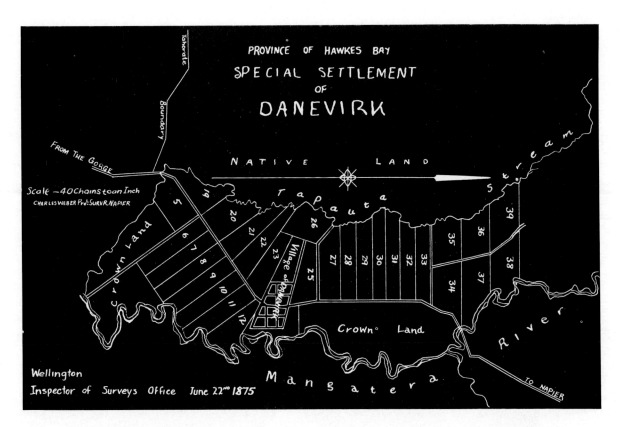

The special settlement "Danevirke", Hawke's Bay Province, 1875.

One danger of the bush-burn was that, through a wind change, it might veer round and destroy homes and other developed properties. Furniture and kitchen gear was often buried in advance for safety, while windows and doors would also be removed. Changing wind conditions during burning posed a constant threat.

New Plymouth

By 1851 New Plymouth had developed a locally farmed area to the extent that the surrounding bush allowed, and a significant trade in agricultural products had been initiated with other coastal settlements. Wheat was the principal crop grown, for in spite of the area of sown pasture, livestock were only of limited consequence. By the beginning of the 1850s, however, with the prospects of continued cropping threatened by declining soil fertility and reduced yields, it was increasingly apparent that the rearing of livestock would probably prove more profitable. Consequently English sown grasses replaced the former grain fields in the district.

By 1850 New Plymouth settlers had built up a considerable local trade in agricultural produce. Although flour was then the single most important export, representing about 70% of the total value, a future trend in Taranaki production was beginning to emerge: the second export by value was butter, representing 12% of the total value of exports.

The Richmond-Atkinson family correspondence contains numerous references which clearly show that control of bush-burn was not the only inherent problem in clearing the land:

> ... the bush party returned earlier than usual. There had nearly been a sad accident in felling a pine; in falling it struck another tree and the head split ... Hal and Arthur were quite buried and for a minute James and Charlie feared they were killed or much injured but thank God they were quite unhurt....
> (JMR)

Surveyor's camp, Hawera, circa 1870.

In November 1854 Harry Atkinson writes of bush clearing and tree felling:

> ... We felled 28 in our clearing, the largest measured 21 feet 11 inches the smallest 6 feet. ... The fire if it is a good burn destroys all the small stuff, leaves, and burns the surface of the ground as well.

Reducing all the vegetation to ash was often referred to as a white burn. Clearing up for further burning after only a partial burn was, as noted, most unpleasant for all.

Bush Farm Cheese- and-Butter Making — the trade in farm dairy produce near New Plymouth by the 1850s

The following extracts from the Richmond-Atkinson letters and diaries suggests a fairly active programme of butter-and cheese-making on farms near New Plymouth and highlights farmer involvement in the trade of dairy produce using the barter system. The potential Australian market was viewed with much interest. By the late 1850s the disposal of surplus dairy produce was well established.

31 July (1854)

... We have had a splendid winter.... There is very little milking to do, all our cows but one have been sent back to the bush as we have not enough grass for them and our small crop of mangel wurzel ... was burnt during the great fires last summer.

10 November (1854)

... Cattle do well in the bush but they require a clearing to come out into to sun themselves.... We have no fences up so they go where they like but they are never away any long time....

18 April (1855)

Harry's agricultural or rather pastoral pursuits bring in very little money at present. But the livestock is increasing ... and the land itself has been doubled in value by the opening of the road....

2 December (1855)

Fri. Dec. 7 ... Our bush cattle-keeping has been on the whole a poor spec for us. We have not only lost much time in hunting the things, but have also lost several animals.... The English grasses are shamefully full of weeds amongst other things docks. The bag marked *Festuca pratensis* appears to be ryegrass. T King advised me against sowing any of it for seed, but WSA sowed some and has a fine patch of almost pure ryegrass. The *Alopecurus* contains about a twentieth of the genuine grass.... The Timothy has come up well and looks promising.... Anything interesting in the seed line will always be acceptable....

25 August (1856)

The large exports to Australia and the considerable addition to our population have raised the necessities of life to famine price ... Harry got £20 on Saturday for a cow which cost him £11 about 6 months ago.

7 December (1856)

Cal and Charlie are working for me very briskly. The cowshed which you saw in a halfbuilt state is now a flourishing workshop....

31 December (1856)

... We propose taking to cheese-making as our profession. Harry is to grow the milk for us so that we shall only have to change it into cheese and not be bothered with milking cows.

4 January (1857)

We shall not be settled for some time yet ... we intend taking up cheese-making and that is the real reason of Ar's wishing for the du Martin house; 26 feet by 14 feet and a lean to is rather an extensive addition for dairy purposes you may think .... Ar would never undertake such a business if it involved keeping many extra people in the house for the milking and single-handed he could not milk enough cows to make it pay.

3 May (1857)

We shall probably not get more for our cheese than we gave Harry for his milk. Virtue may be its own reward, but cheese-making is not.

20 June (1857)

Feb. 16 I attended a general meeting of the men of Hurworth in the evening to discuss the prospects of cheese-making and the advisability of opening communication with Sydney and Melbourne. We are thinking of setting up a shop in the town (for the sale of pork, cheese, ham, bacon, butter, mutton, eggs, fowls, etc). Hugh Reynolds to be shopkeeper.

Mar. 16 Turned cheeses, cut vinegar barrel in two for pickle tubs, put pork into salt.... Adams is going to Sydney.... We have sold Adams 1 cwt of our first cheeses at 11d per lb — this is the first cheese we have sold.

30 June (1857)

I am at last in my shop selling cabbages, butter, cheese, etc. It is the only shop of the sort in town and predicted by all to do well. The Atkinsons, Richmonds and ourselves are shareholders.... Farmers here if they want to get rid of their produce must barter it with the store-keepers....

15 November (1857)
... a great deal of our business is done by barter (we exchange cheese for milk, butter and eggs) ... we have ham and bacon for our own curing.

6 December (1857)
Harry has bought all James' pigs.... How a butter- or cheese-making dairy farm is to pay without pigs I am at a loss to imagine....

Wairarapa Small Farm Settlement

In the 1850s bush dairying in the Wairarapa was closely associated with the establishment and later expansion of smallholder farm settlements. Wairarapa was essentially a spillover of the New Zealand Company's Port Nicholson settlement. The early colonists, both capitalists and agricultural labourers, had arrived in Port Nicholson to find that not only did their prospective sections still await survey, but also that the validity of their purchases was still in some doubt.

Lack of both land and capital resulted in special hardships for agricultural labourers and artisans, since their living depended on a cash income obtained by working for those colonists who had land and capital

A Manawatu bush settler's "dream"!

and could thus afford to employ them. The real problem, however, was that there were more men seeking work on the land than there were jobs available or land that could be worked.

Such labour and land problems in the Wellington area attracted those with capital to the Wairarapa. Consequently, in the 1840s, despite access and tenure difficulties pastoralists occupied much of the eastern grass and scrublands until by the middle of 1850 only the more heavily forested areas of the Wairarapa remained unoccupied and available for settlement.

Because of earlier limited success in the lands around Wellington and the Hutt Valley, the potential of small farming in the Wairarapa bushland was soon appreciated. The 1853 purchase of most of the Wairarapa land and the reduced price of much rural land initiated a trend towards the expansion of closer, small farm settlement. The Wairarapa Small Farmers' Association was founded and it negotitated the purchase of 25,000 acres, but such was the power of the earlier pastoral lobby and its continuing influence that most of the land left for purchase was in the bush. The successful settlers were largely farm labourers from the New Zealand Company's Wellington settlement. The proposed 40-acre bush plots posed problems since clearing such a large area was a major exercise. Later, however, the same size would prove inadequate to support a settler's family. Despite such experience with private small farmer settlements, it became government policy to encourage settlement on even smaller plots of land. With the growing need for land, small farm settlement sites were dispersed over four subdivisions in Masterton, Greytown, Taratahi and Moroa, beginning a new era in the Wairarapa.

Although Featherston had no organised small farm settlement comparable with those further north, 10,000 acres were subdivided by the provincial government for small farm settlement. The sections were mostly between 5 and 50 acres. Land prices were dearer in Featherston than in earlier settlements, but the settlers there had a locational advantage over small farmers to the north for they were 8 miles nearer the

Worksop farm 1869. This farmhouse became the first accommodation house in Masterton and also the terminus for the Wellington-Masterton coach service.

Wellington market. It was partly for this reason that early in the 1880s the Featherston farmers became the first in the Wairarapa to experiment in commercial dairying and factory production for export.

The new immigrants were settled on 10-acre sections in the bush area, mainly so that they could carry out road construction. Farm work on their small properties was initially a subsidiary activity. Permanently settled immigrants close to the areas of proposed road construction were regarded as one of the most reliable and readily available pools of labour for such work. By this time, small farm settlement was well recognised officially as a very useful means of carrying out bush clearing and land development.

Many moving from the Wellington area to the Wairarapa small farm settlements in 1854 already had local farming experience and were well aware of the local market situation for dairy produce, and most notably the expanding Australian markets occasioned by the increasing gold rush population. Such optimism over future prospects, however, had to be immediately qualified because of the difficulties and costs of transporting farm produce over the Rimutaka Range. Furthermore, the economic situation changed, the Australian gold rushes neither generated nor maintained a stable market demand. Thus Wellington merchants in 1855 and 1856 complained of "the dullness of business", while one commented that he did not expect a pound of New Zealand butter would be wanted in Australia within a few years.

This early depression struck at both farm communities in the Wairarapa. Many sheep farming pastoralists depended on dairying as an immediate cash crop. While pastoralists could shift to other forms of production the bush small farmer had no such alternatives. Whatever the state of the current market for dairy produce, the small farmer, with no viable alternative, was compelled to farm on a less and less remunerative basis. Thus dairy cows and milk production continued in the western Wairarapa while eastern runholders persisted with wool and beef on their extensive stations.

Isolation from the Wellington market was a particular difficulty. The first agricultural produce was transported over the Rimutaka Hill in 1850 by one of the pastoralists. As late as 1853 the Wairarapa side of the hill route was still a bridle track and until 1856 was too narrow to accommodate wheeled vehicles. The opening of the Rimutaka Road to wheeled traffic did not, however, mark the end of communication problems with Wellington. A further 2 years were to pass before a cart road to Masterton became practicable so road building continued to be critical in providing improved access.

During this early period butter was usually the only dairy product with a marketable value but usually it was bartered rather than sold for cash. The *Wairarapa Standard* of 15 November 1886 reported that this practice was even then still widespread. Surplus farmhouse butter was taken to the local store where goods and provisions were bartered in return, normally ". . . the storekeeper paying a very nominal price for the butter and as a rule charging a good price for his goods, thus obtaining a profit in two ways". In spite of meagre returns, few Wairarapa farmers tried for alternative dairy products or markets. It may have been that ignorance of cheese-making skills meant that butter was easier to make. It would be difficult, for instance, to produce rennet without slaughter calves, which were highly valued as the basis of the small farmer's future livelihood.

A correspondent from Wellington, writing in the *Wairarapa Standard* of 31 August 1882, asks why the Wairarapa small farmers had been content over the years to receive only 4d/lb for farmhouse butter at the

Opposite page: Wellington 1841.
Watercolour by Charles Heaphy.

Tinakori Road, Wellington, 1840s. Engraving by SC Brees.

local store during the season, while there was available nearby in Wellington a regular market for good salt butter in winter? Under such market conditions salt butter sent to Wellington would have given much better returns. The writer suggests that perhaps such available opportunities were not fully recognised, certainly they were not taken advantage of at the time. Any exporting activity was essentially left to the entrepreneurially inclined merchant to whom small farmers sold their butter for milling, or blending, and then packaging for further sales.

Squatters and Runholders in the Wairarapa: 1840s - 1850s

The Wairarapa was also the scene of some of the first efforts and successes in pastoral farming in New Zealand's tussock grasslands. Most runs included some small-scale cattle raising along with their predominantly sheep-based enterprises. By March 1847, Weld and Bidwell at Wharekaka were running 13 cattle with 3,200 sheep. At Kopungarara, Bidwell grazed about 190 cattle belonging to Ludlam and other Hutt Valley settlers, along with 350 sheep of his own. In 1849 Smith and Revans at Ruangura ran 150 cattle with 2,000 sheep. By 1847 there were 16 runs or cattle stations in the Wairarapa with an estimated 13,000 sheep and 1,400 cattle. In 1849 Allon estimated a total of 20,000 sheep and 2,000 cattle, rising to 37,000 sheep and 3,000 cattle by 1853.

Wool, butter, cheese, pigs and vegetables were shipped to Wellington by a small fleet of coastal schooners. Wellington's *Independent* newspaper, on 24 January 1852, reported the arrival of the schooner *Sea Belle* with a cargo including 40 bales of wool, while a few days later the *Twins* was reported with a cargo that included 41 bales and 3 bags of wool, 10 kegs of butter, 8 hams, as well as onions, bacon and 5 live pigs. As the coastal trade improved there were regular schooner, boat and canoe services along the coast. A road over the Rimutaka Range to Pakuratahi and the Hutt Valley was completed by 1848.

Cattle breeds on the runs were not highly specialised, the general aim being to produce a good, all-purpose animal, and Wairarapa cattle were generally good quality shorthorns. Durhams, a shorthorned breed of New South Wales origin, were common, as were Red Devons bred from Ludlam's valley herd and yielding good beef as well as a reasonable quantity of milk. Wairarapa stock usually came from or through Wellington. Most imported cattle came from Australia, particularly the ports and districts around Sydney and Twofold Bay.

There was abundant, year round pasture for cattle to graze. Even in summer when the tussock was often dry and somewhat unpalatable, there were still taprooted aromatic plants to support stock, and palatable plant species in the swamps. The fern and bushland provided additional forage, especially for cattle, which trampled and opened up scrub and fernland for other stock. There was further fodder for cattle in the bush and forest for they ate the lower leaves of broad-leaved trees as well as small shrubs and plants. Cattle also had the advantage of being able to graze a much wider range of fodder plants than sheep. Further grass growth was never checked sufficiently to affect stock condition seriously, while fodder was so readily available for the stock themselves that it saved both labour and capital since cattle were never housed or stalled but left outdoors for year round grazing.

3

Subsistence

AND SELF-SUFFICIENCY

In the South Island settlements in Nelson, Canterbury and Otago, as in the settlements of the North Island, agriculture and livestock production had made some limited but uneven progress. Small farms on former bush or fern covered lands were generally located in and around the newly established settlements, the Waimea in Nelson, Banks Peninsula and Riccarton in Canterbury, and the Otago Peninsula and the Taieri in Otago.

The earliest European settlements, the whaling stations, scattered along the eastern and southern coasts of the South Island, had little, if any, real contact with the other European settlements of the time and witnessed little permanent agricultural and pastoral development. One of the few exceptions was Johnny Jones' whaling station, later a farm settlement, at Waikouaiti which, with the arrival of the first settlers in Otago in 1848, became a major early source of food supply.

Early cattle imports were recorded from Australia in the 1830s and 1840s for Nelson, Canterbury, Akaroa, Waikouaiti, and Otago, and continued throughout the 1850s with further smaller consignments from Britain as well. Regular trade in cattle took place in the settlements, as is evidenced in the earliest newspapers, and stock were often imported on consignment.

Squatters and Runholders in Nelson and Marlborough

In the Wairarapa, early dairy development had been associated primarily with small farm settlers and to a lesser extent with squatter runholders. Both were motivated by the need for self-sufficiency in food supply as well as by the commercial opportunities offered by the nearby, expanding Wellington market. By contrast, in Nelson and Marlborough after the early failures of the first smallholder settlers, developments in dairying were associated with bigger holdings.

When the selected site of the New Zealand Company's Nelson settlement was found to have insufficient good quality agricultural land to honour earlier promises to immigrants, the primary objective in opening up the back country regions was a search for extensive, and accessible grazing lands for the growing sheep numbers. Apart from the continuing search for gold, it was the discovery of a sheep-droving route from Nelson to Canterbury over the old Tophouse Pass track to the Wairau and Buller Valleys that was to prove decisive in setting the patterns of early settlement and development of land use in districts like Wairau or back

Early settler's tent and first timber hut.

country Nelson.

Family correspondence of the Hon Constantine Dillon records many interesting details of these developments. Dillon was a landowner, run-holder and squatter. Before arriving in Nelson in 1842, at the age of 27 years, he had served in the army in England, was aide-de-camp to the Lord Lieutenant of Ireland, and was one of Lord Durham's secretaries in Canada in 1838. It was while serving with Durham that Dillon made the acquaintance of Edward Gibbon Wakefield, who may have directed his attention to New Zealand as a land full of promise and opportunities. Dillon was persuaded to invest in several New Zealand Company land orders.

The grazing lands of the pioneer Nelson settlement extended no further west than the Waimeas when Dillon and a partner first began grazing stock in the Upper Motueka Valley in 1843. The annual 1845 stock returns recorded that both had 750 sheep, while in the same year the local agent of the New Zealand Company allocated them "depasturage licences" for that land at £10 a year.

Dillon's letters describe plans and preparations and the running of various properties.

16 Jan 1843 Nelson
... We came out at once to the country and settled on one of our suburban sections on the plains of Waimea about ten miles from Nelson but navigable up the river and we can get up to within three miles of our house by water.

The first cow and bull. Sketch by Charles Heaphy.

... The plain on which we are is for miles around all in fern, about shoulders high in some places, and flax in others. Some of the fern in the woods is very pretty ... but the roots are very thickly matted together, for which reason I have advised Henry Story, if he comes, to bring strong, heavy harrows and rakes. The fern was all burnt off by the surveyors so that it is not higher now than one's knees but where some escaped it is up to one's shoulders.

The land is generally very good since all the farming people agree it is fine. It is an upper soil, apparently of decomposed vegetable matter about a foot deep, with a sub-soil, varying from 2 -3 feet in depth, of sandy clay. The land is not generally quite so good as woodland but it can be cleared and ploughed and the seed put in properly for between £5 and £6 an acre, which is what it would cost in England to clear land which has got foul. The woodland costs about £30 an acre to clear ....

I have got some cows and bullocks but I have not seen them for some time, as there is very fine forage for them all about the country and they are turned out to shift for themselves. Keeping cows will answer, if the price of butter remains as it is at 3 shillings a pound. However, none of my cows give me any milk yet ... but I hope they will soon calve.

10 April 1844

... I have got my 50 acre section broken up and fenced in. Half was in corn last summer. I had 8 acres of barley. Our cattle cost nothing to keep as they run all over the plain but they cost us a great deal of time to find and we lose all the manure. But this we hope to remedy as soon as we can grow food for them and keep them in the yards ... I have 12 working bullocks ... Milking cows nine, in milk at present four. Average butter per cow last summer, 5½ lbs a week, price of butter from 1/9d to 2/- lb. The cows, like the working cattle, graze on the plain and are brought in morning and evening to be milked.

Garden one acre.... My crops this season will be 20 acres of wheat, 6 of oats, 8 of barley, 4 of grass, 10 potatoes.

... There is an immense preponderance of labour and yet good labour is dear ... we have an immense number of mechanics instead of agricultural labourers.

... Imported provisions are wonderfully cheap when one buys a whole keg or package ... I am selling cows for £12, steers £17 ... fresh beef and mutton 10d lb. We shall soon be able to sell it for 3d or 6d a lb and then it will be cheaper than salt flesh.... My outer fences are post-and-rails, cut in the forest of a tree called totara, and the inner ones of what is called in England staple-and-board. These have all been done by contract, both cutting and putting up, which is a much cheaper way of doing things than doing work and carting the stuff. My outer fences cost me, ... including carting £78. As the stuff had to be brought from five miles off it took 180 days to bring it, so that the cart had to go 1,800 miles for the stuff. Everything is getting cheap now, roads opened in the woods, which assist very much. Iron hurdles would be very expensive to bring out and not better than what I can get made here for 1/2d a piece.

... Mr Duppa on the other side of the Waimea River... is my partner in a herd of cattle. He has very large herds and a large dairy where he makes a great deal of butter and cheese. His dairy is of about 40 cows.

George Duppa's farm on the Waimea Plains, 1840s.

In an 1848 ballot, Dillon drew a 200 acre rural section in the Wairau. In the same year, less than 3 months after acquiring the property, Dillon was appointed to the position of Civil and Military Secretary by Sir George Grey, which resulted in an inevitable move to Auckland. A later letter tells of the various plans prepared for running the farms in view of the forthcoming move.

1 May 1848. Nelson.
... I have been very busy making arrangement for my stock.... Dr Monro undertakes my sheep, 2,000 in number. He takes all the expenses upon himself for which he is to have one-third of the wool and one-fifth of the

increase in about ten months, that is when my lambing is over. I shall have about 2,800 sheep, which is a very tidy flock for this Island, where sheep are on an average worth £1 a head. My cattle and horses, 60 head of the former...I have settled with a man who is to take up a dairy which I have in the Wairau district. He is to find all expenses also and is to be paid in produce, that is, he is to have two-thirds of the butter and cheese, and five percent of the cattle reared and that reach the age of six months but I am to build the dairy, house, yards, sheds, find all the dairy implements, in fact to start the whole thing and to deliver him the cattle on the spot.

My house is to be let this year to a Mr Duppa for the ridiculously small sum of £30, including a garden and ten acres of grass paddocks. My farm is taken by a man who gives me one-half of all it produces in kind, finding agricultural implements and the seed and labour. This seems a funny arrangement but everything is to be had here but money. I shall therefore have my flour and oats sent to me at Auckland, as well as butter and cheese from my dairy, and bacon.

The house and dairy set up at Wairau cost about £80 or £90. In a letter to his wife in 1850, after he had made his first visit from Auckland to his Wairau estates, Dillon gives a detailed account of his dairy operations.

6 December 1850

... The dairy is very well managed. Brydon and his wife are very active and good servants and as Cautly lives here he is able to see whether they do justice or not and he says they do ample justice to the cows, indeed I have seen so myself the last ten days.

The cattle, of which there will soon be 100 head, are disgustingly fat, indeed the cows too much so for milking. I shall bring back with me a keg of 56lbs of butter. There is no fresh butter made here. It is all made salt and put into kegs at once. In about a month from this date they will have 27 or 28 cows in milk and then they will make nothing but cheese. There is no cheese now ripe enough to take away. I am going to Nelson on Thursday with Cautly when we shall drive about 6 fat bullocks to that place with us. Fat cattle fetch as high a price now in Nelson as ever they fetched.

... The cattle in the Wairau are fast dying from starvation. It is so overstocked ... people are all anxious to sell their stock there. They say they must raise artificial pasture. Cautly and I are going to look at the Wairau farm with a view to laying it down to grass....

On the whole, however, I am very much satisfied with the dairy and cattle and I am sure it will be an excellent speculation. The house and the dairy are both very substantial buildings. The house has three rooms and a kitchen and loft, all good-sized lofty rooms. One is called my room. The dairy is capacious and well ventilated with a good cheese room attached. It is built very strongly of clay walls 18 inches thick and good stout timbers in it. The stockyard and milking yards are better than the one we had at Waimea.

Cautly is now enclosing a piece of 35 or 40 acres to keep cows in a few days before they calve as they are apt to stow themselves away then.

The boar was killed by a dog at the mouth of the river so that the pigs have not increased but I am going to make provision for replacing them. Next winter they will make some bacon. There are now at the dairy five very nice pigs preparing for that fate.

By the early 1850s there were several dairies in the Wairau making cheese and salted down butter for the Wellington and Nelson markets. The lower Wairau Plain seemed fit for little other than grazing cattle, so

Opposite page above:
Pigeon Bay in the early 1880s and the Hay family homestead.

Opposite page below:
Deans farm, Riccarton. Sketch by WBD Mantell.

An early settler's "dream" of country life. Such primitive conditions had to be endured while settlers established themselves on the land.

Above: Holmes Bay. The original homestead was later destroyed by fire.

A tent bush dwelling, West Coast.

that several of the pastoralists who were also landowners used their allotments for this purpose. Raising cattle for the butcher, however, was the farmer's prime objective.

Smallholder Immigrant Settlers in the 1840s in Nelson

Earliest European settlement dated back to the early 1840s and reports in the *Nelson Examiner* refer to cattle imports destined for both small subsistence holdings and larger cattle and sheep stations. The comments of at least one early Nelson settler would suggest, however, that the acquisition of dairy cows could prove extremely costly for many smallholders or farm labourers, with a working bullock worth more than £30 and a cow as much as £50! In spite of such imports it would appear that cattle, as well as dairy produce, were seldom in truly plentiful supply. In both 1842 and 1843, for instance, advertisements in the local newspaper offered for sale kegs of imported Irish butter and preserved beef.

By 1851, after a somewhat disastrous beginning, the settlement of Nelson had become one of the foremost cropping areas in the Colony, with almost one-third of the total crop land. The initial Nelson settlement had been characterised by a large number of extremely small holdings, many of which were only 4-8 acres in size. As in other early settlements, cattle would have been depastured on the nearby surrounding bush and scrub and generally owned in small numbers.

Like settlers elsewhere, those in Nelson had set out to recreate the traditional mixed farming crop and livestock economy with which they were familiar, and cattle were well integrated into such farming.

Early Canterbury and the Deans Brothers at Riccarton

The Deans family made a major contribution to the development of farming in Canterbury and particularly to the beginnings of cattle and dairy farming in and around early Christchurch before the arrival of the "first four ships" and the first settlers. In 1842 William Deans and his family, accompanied by two Ayreshire farm labourers, Gebbie and Manson, sailed from Wellington to Banks Peninsula in company with Ebenezer Hay. A year later John Deans sailed from Sydney to Port Cooper to join the family. Accompanied by his farm workers, William Deans moved from Port Levy to Sumner, and thence up the Avon River to Riccarton where a slab hut and sheep and cattle stockyards were soon built. He also began a dairying enterprise with Gebbie and Manson near Gebbies Flat, a convenient location for servicing the whaling vessels and also supplying the later Wellington and Australian markets.

Ebenezer Hay on the voyage with Deans, had been duly impressed with Banks Peninsula. A year later, along with Captain Sinclair and their respective households (15 people, two cows and a calf) he settled in Pigeon Bay, about 8 miles from Port Cooper. While the Hay family occupied the flat at the head of the main bay the Sinclair family settled in Holmes Bay to the west. These two families and the Deans were followed by the Greenwood family from New South Wales who settled at Purau. All four family groups were engaged in stock farming rather than cropping. The Deans brothers carried out a little cropping, but grain markets were generally too far away and transport too costly. By February 1844 it is recorded they had a well-established stock farm with 76 head of cattle, 3 horses and 50 sheep.

Although sheep numbers were growing rapidly they were regarded as an investment for the future rather than a major source of immediate income. Cattle on the other hand, provided not only prospects for a future trade in beef animals but also an immediate return from dairying. For example, William Deans at Riccarton, immediately built a cowshed with 10 double bails and set about importing cattle and sheep from Australia, and at the end of the first season sent a consignment of cheese to Wellington. Cheese-making was, in fact, one of the main occupations of these first Canterbury stock farmers, not only at Riccarton, Pigeon Bay and Port Cooper, but also at Rhodes' settlement in Akaroa. There was some trade in salted butter, mainly to the whaling vessels visiting Akaroa. A trade in beef cattle developed first from Rhodes' station, but by 1845 all were beginning to send fat cattle to the Wellington market, a 6 day voyage.

By 1846 the Deans' herd had increased to a total of 130 head of cattle, and the market for fat stock was growing, especially to feed troops garrisoned in Wellington. Until then the Riccarton stock farmers had concentrated upon Durham cattle for their good milking qualities, but in August 1848 John Deans commented in a letter, "... Our present stock of cattle are good enough to lock up as good milkers but they are nothing like some of the breeds out here for the butchers — they grow fast enough and fat enough, but the beef is coarse on the grain, and generally gets dry after being in the salt some time".

By the time the advance surveyors for the Canterbury settlement arrived the Deans had a good stock of capital, as well as cattle and good pastures for grazing and fattening. By burning and grubbing up tussock and flax, and then ploughing and sowing English grasses they had achieved high carrying capacities and improved paddocks.

Apart from the efforts of the early pioneer families, Deans, Hay, Sinclair and Gebbie, by 1851 the Canterbury settlement, having been founded only a year or two earlier, had experienced minimal agricultural development. The original French settlers in Akaroa who arrived in 1840, occupied small bush-cleared holdings of about 5 acres. All of these farms were subsistence in nature and had few if any livestock.

The earliest Canterbury and Banks Peninsula farming was mainly concerned with livestock; in January 1849, the Deans on their Riccarton farm were running about 150 cattle and 1,000 sheep. But apart from some early dairy enterprises a different direction was emerging in Canterbury farming. By 1851, 11 sheep runs had been established in the

Taieri Plains, 1860s.

province, running about 28,400 sheep or approximately 12% of New Zealand's sheep population.

Typical country settlement, Otago, 1870s.

Early Settlers in the Otago Block

Dairying in Otago pre-dated the settlement of the Otago Block in the late 1840s, for in 1838 Johnny Jones had acquired the whaling station at Waikouaiti, some miles north of the Otago Peninsula. He purchased more land in 1840 to settle several farming families and so extended the activities on his holdings.

In 1844 one of the first attempts at permanent and full-time farming on the Otago Peninsula was made by two squatters, Anderson and Rowen. The land was low-lying, sandy and bush clad. Areas were slowly cleared for pastures, livestock was introduced from Wellington, and Kelvin Grove farm was established, in time carrying 500-600 sheep and 70 cattle. Because of insufficient open land, too much bush cover and drifting sand, the land proved unsuitable for pastoral and stock farming so the herd was divided and the two men left Otakou.

In 1844 Tuckett recorded cattle at various places, among them Riverton and most notably Moeraki at Waikouaiti where the original imports had increased to about 200 head. The first ships of the Otago Block settlement carried no livestock with them but settlers soon acquired cattle from the original stock of Johnny Jones at Waikouaiti. Cattle were generally grazed in the surrounding bush and scrub in the nearby Taieri and Otago Peninsula.

Otago's early settlers were most familiar with the traditional Ayrshire breed and the Scottish colonists introduced them into the new settlement as soon as the necessary arrangements could be made. The first Ayrshire bull was reported to have arrived on the *Philip Laing* in 1848, consigned to the first Presbyterian minister, the Rev Dr Burns, and was followed by further imports for the settlers on later vessels.

Within a year of the founding of the Otago settlement farming had developed on both sides of the harbour, as well as in the nearby

View of the Otago Peninsula.

Taieri-Tokomairiro and Clutha areas, with settlers widely scattered throughout the Otago block. Most of the cattle were to be found in the more intensively farmed wooded lands around the upper harbour, stretching over the rolling hills towards the Peninsula and the Taieri Plains. The small mixed farms were little more than semi-subsistence holdings of about 10 acres located in small clearings in the bush.

4

FARMHOUSE DAIRYING

The pioneer settler, confined by the limited acreage of the bush or fern covered holdings and struggling to develop an economically viable and productive farm unit, was constantly frustrated by the limited production and marketing opportunities available to him in the young Colony. The size of the local market for dairy produce was governed by the growth of population and settlement, but the communication and transport systems were too primitive to link up the scattered and isolated settlements and this effectively restricted any market to just the immediate locality.

The 1840s saw a considerable influx of European settlers and with large scale colonisation and settlement a rudimentary form of dairying and dairy processing developed. However, dairying was not primarily undertaken as a commercial enterprise, rather it was a contribution to the settler's subsistence and self-sufficiency on his new holding. Even on some town allotments settlers often ran one or two cows to meet family requirements, leaving the livestock to browse freely wherever they chose.

Generally the quantity of dairy produce was sufficient to meet the needs of the settler and his family. As time went on dairy produce surpluses grew, particularly in those periods when the market prices were such as to encourage the production of butter and cheese. When prices were less favourable the cattle, instead of being milked, could be sold for their meat. Thus, until refrigeration made export trade in butter possible the Jersey was never as popular as the dual-purpose breeds.

Overall, however, dairy production was discouraged by the lack of outlets which consequently limited possibilities for growth and expansion. The local market would often be glutted and export efforts proved only partially successful. The dairy market in the Colony offered few real prospects in spite of significant population increases.

Milking Practices

Almost without exception cows were milked in the open stockyard or field. They were usually put into bails formed by two upright poles driven into the ground and drawn together over the beast's head to secure the animal during milking. Hand milking was, of course, universal, and was usually the task of the settler's wife and children. Until the introduction of mechanical milking methods, hand milking involved such a significant labour input that a critical condition of expanding a dairy herd was a

A travelling tinker effects repairs while the children of the family look on.

comparable increase in available family labour. Early settlers imagined that dairy cows were highly sensitive to even slight changes in milking routines and farmers cautioned that ". . . if possible the same milker should have the same cows every day, and should keep exactly the same time in milking, and milk them in the same order".

Farmhouse Butter-Making Methods

In the farm dairy the milk was strained through a fine, wire-mesh strainer and left in setting pans for the cream to rise. Shallow, flat containers were recommended and proved to be very efficient. These setting pans were made of a variety of materials but most New Zealand farmers used tinned iron because it was so readily available. Great care had to be taken with the pans, the metal seams were easily clogged and fouled by milk residue which, if not immediately removed, tainted the contents of the next batch of milk.

Throughout the setting process the dairy had to be kept as cool as possible and was designed and constructed accordingly. Usually 24 hours was allowed for the cream to set, although frequently 36 to 48 hours may have been allowed. When pans of milk were placed in cool water, the cream would rise more quickly and setting time was shortened. Once set, the cream was removed with a skimming saucer or ladle and collected in containers where it was left until ready for churning which was also done by hand. Because of the extremely small size of so many dairy herds churning was usually done only three or four times a week.

Once a small quantity of butter had been made it was usually packed in brine in well seasoned wooden kegs, which were frequently imported

"A New Zealand milkmaid".
Photo, C Fitzgerald.

although some were manufactured locally of pine or tawa. The latter, however, were often criticised because of finishing and bonding with rusty iron hoops. Furthermore, such kegs were sometimes unsound so that the brine leaked out and the contents consequently deteriorated. On such occasions the butter was sold on the local market at lower prices. Some butter-makers favoured the use of dry salt on top of the butter in the kegs as a further preservative, but this practice rarely proved satisfactory, as evidenced by the poor condition of the contents when the keg was opened for use.

Many settlers potted down quantitites of butter in earthenware vessels or small tin cylinders in order to have supplies available for winter sales or for their own consumption. While such means of preservation were adequate for short-term storage, the butter tended to deteriorate badly if it was kept for any length of time. Butter was wrapped in a variety of materials, linen and calico being the most frequently used, although the use of dock or cabbage leaves was not unknown. (In 1853 the process of making vegetable parchment was invented.)

Traditional Farmhouse Cheese-Making Practices

Methods of cheese-making have changed little with time, except that science and technology have improved many of the implements and much of the plant used. The only methods available were those passed down from generation to generation. Thus the early settlers brought their own traditions of cheese-making: Scottish "Dunlop" at Highcliff for the Mathiesons or Somerset "Cheddar" at Flemington for the Hardings. Dairymaids and farmers' wives had all learned cheese-making largely by experience, realising that certain methods of handling milk produced better results.

Once collected, milk was left to stand for the cream to rise and the warm milk to ripen naturally. Rennet extract from the linings of calf stomachs was added and the warm milk was placed in large, round iron tubs to bring about coagulation and form the rennet curd. Although not clearly understood, the basic factor throughout the cheese-making process was "the acid to moisture balance".

The curd was left till fairly firm before it was drained. Later it was torn or broken up, usually by hand but sometimes with crude utensils. The small pieces of curd and the whey were heated and stirred continuously, and removal of the large bulk of the whey by dipping off proved a tedious process. The next stage was the drainage and ripening of the curd, for on this depended the ultimate quality of the cheese in terms of balance, body, and texture. The curd was then again torn up or cut into small pieces, although later primitive cheese mills evolved to tear the curd in hand turned, spiked rollers. Salting was the next essential process to flavour and preserve the curd. Curd was either salted before or after pressing, alternatively unsalted curd was pressed in moulds or hoops and the cheese subsequently "pickled" in brine. The mould or hoop was then filled and pressed with primitive, dead weight presses. The cheeses were constantly removed for scraping, salting, dressing and rewrapping in cloths, while the pressing itself ensured consolidation into a cheese. Finally the cheese ripened on shelves in the store, a process which demanded constant attention.

Imports and Exports of Butter and Cheese

Although the amount of cheese and butter produced increased, there were still some imports from Britain and Australia to meet the growing local demand as the population expanded. There were no regular import patterns, amounts varied greatly from year to year according to market circumstances. Cheese imports ranged from a maximum of 1,852 cwt in 1868 to a mere 23 cwt in 1872. As a rule New Zealand made cheeses retailed at a lower price than imported varieties. In the period 1868-1881, 50% of cheese imports came from Britain but by the early 1870s the Australian state of Victoria supplied nearly 60% of all imported cheese.

Butter imports also varied, from a maximum in 1868 of 5,132 cwt to a low in 1877 of a mere 21 cwt. Butter prices reflected a less clear difference between imported and local produce. Once again Victoria dominated the butter trade between 1868 and 1881, accounting for nearly 75% of all imports.

Throughout the period 1853-81 dairy exports were insignificant, the highest percentages of dairy produce in 1853 and 1855 accounted for only 3.8% of total exports by value, and in the 1860s and later, seldom exceeded 1%. This situation did not necessarily indicate a lack of effort or enthusiasm in trying to develop an export trade but rather it highlighted both the isolated location and the lack of technological expertise associated with the preservation of quality in perishable exports which had to undergo long sea voyages. Many efforts were made to develop effective storage processes and preservation techniques but all apparently failed. The tinning of butter for export was one approach tried but it failed because of imperfect understanding of the role of bacteria in food preservation and the need for sterilisation. Butter was also packed in brine, although the use of salt alone was not practicable because of the very high costs involved. Butter exports suffered from the tendency to

Above: Outdoor milking,
Paeroa, 1902.
Photo A Sherlock.

Left: Feeding a calf.

deteriorate rapidly to the consistency of axle grease and to develop a "fishy" flavour due to the blending or mixing of different farm butters for export. Cheese did not deteriorate so rapidly but it also posed problems. In one recorded instance, prime quality cheese was exported to London in strong cases filled with kiln-dried oatseed, but when it arrived a month or two later it had perished. This was the fate of most exported dairy produce until the development of refrigeration.

As with imports there was no regular pattern of dairy exports. In 1861 only 26 cwt of butter was exported and in 1866 only 13 cwt of cheese, but 10 years later in 1877 there was a maximum export of both dairy commodities, 4,999 cwt of cheese and 5,286 cwt of butter. During the period 1868-1881 Australia was the main market. Throughout the pre-refrigeration period there was a continuing interaction in the trade in dairy produce between Victoria and New Zealand. After the gold rushes, when local markets tended to grow rapidly to displace exports, there were efforts to re-establish an export butter and cheese trade but the quantities involved were neglible.

One paradox was the success of imported butter and cheese, all the evidence suggests that Irish butter and English cheese arrived in New Zealand in good condition. A contributing factor might have been that the Irish were traditionally skilled butter-makers who packed only the highest quality produce under the best available conditions. The bulk of the trade would also have been in summer made butter, butter made at the best time of the year and under the most favourable conditions. The butter was well packed in oak firkins and heavily salted, as well as packed

Above: Nellie prepares to milk "Clover".

Opposite page: A moment's repose in the dairy.
Photo John Webster.

Milking, note the cow's head is firmly fixed in the bail.

in salt to preserve it. A brand selling locally as Double Rose Cork proved to be one of the most popular imported butters on the New Zealand market.

"Port Cooper" and "Akaroa" Cheese and the Early Trans-Tasman Dairy Trade

In November 1839 Captain WB Rhodes landed 30 to 40 shorthorn cattle, including two bulls, at £16 per head. These dairy cattle, "nearly pure Durham", came from the Hunter River district of New South Wales and were put ashore at Flea Bay in Akaroa Harbour, where they were left in the charge of William Green and Thomas Geed. The open land near Akaroa Heads provided good grazing for the cattle. Rhodes himself did not remain long, the Akaroa cattle station being only one of several of his ventures in New Zealand.

Early French settlers in Akaroa brought no cattle with them, nor could they afford Rhodes' cattle when the latter were put up for sale. However, the young settlement provided a ready market for some of the early dairies that began to develop around Banks Peninsula. After 1843, for example, butter made in a Pigeon Bay dairy sold for as much as two or three shillings a pound, depending on quality and quantity. Hay, a member of one of the pioneering families in Pigeon Bay, was reported to have walked over the hills to the French settlement once a week carrying a pack containing 30 to 70 lbs of butter. It was a fatiguing journey of 30 miles over steep and rugged track and was undertaken once a week for 2 years. Sometimes part of the weekly production was salted and sent to Wellington where there was the possibility of sale to an overseas vessel. From 1850 onwards, after the arrival in Port Cooper of the *Monarch* and the first four ships of the Canterbury settlement, Banks Peninsula settlers found a ready market for farmhouse butter and cheese.

The Hays and the Sinclairs arrived in Pigeon Bay in April 1843. Soon after arriving the Hays sold their schooner to WB Rhodes in return for the purchase of ten of the "horned" cattle landed in Akaroa. The cattle were an excellent buy at £29 a head. There was only a narrow track over the hill through otherwise unbroken forest cover, while a mountain 1,300 feet

Cowshed at Kahouri Bridge, Stratford, 1890s.

high and covered in dense bush lay between Pigeon Bay and Akaroa. The families set to work to widen the track to 6 feet so that driving stock would be possible. Eight experienced men were employed in clearing the track which was completed in 3 weeks, and in a single day all the cattle were brought safely home to Pigeon Bay from Akaroa. Hay and Sinclair each took a half share and began farming in earnest.

In early issues of the *Lyttelton Times* there are repeated references to local dairy farming activities. In both the reports and small advertisements were references indicative of extensive dairying operations; cheese and cheese utensils for sale; butter and butter churns for sale; dairy cattle from Sydney on consignment by local stock agents; land and farm sales; grass seed for sale; farm labour wanted in dairies; details of weekly auctions of livestock; shipping services to the bays; and departures and cargoes.

In addition to the dairies on the Peninsula there were others around Christchurch, and consequently there was a significant annual export of butter and cheese from Canterbury. In 1855, more than 3 tons of butter were exported to the rest of New Zealand, mainly to Wellington, and over 2 tons to Australia, while 22 tons of cheese went to New Zealand markets and 12 tons to Australian markets. During this time dairy produce accounted for 10% of the total value of exports from the province.

5

Assault

ON THE BUSH

Clearing large areas of the dense forest covered bush districts in the North Island for cultivation or surface sowing in pasture, was an arduous process normally requiring years of constant labour and continued perseverance. This work began in the 1850s and 1860s in a small way, in isolated districts. As organised settlement increased bush clearing gathered momentum, a peak being reached in the 1870s and 1880s. Initially the clearing was concentrated on the lowland forested areas but by the turn of the century it had extended to the margins of the hill country.

Work began by cutting down the underscrub. The area would be gone over again, and cuts made half way through the base of the remaining tree-trunks. By felling one tree so that it fell on to others, a large number of trees could be brought down together in what was known as a drive. A strong wind from an appropriate direction at the correct time could assist in the felling process. This work of felling and undercutting was done in the winter months, and then the whole mass of scrub, fern toppings, and trees was left to dry out.

In summer, the tangled mass was set alight and the underscrub, leaves, and branches were soon burnt, leaving only charred black tree-trunks and logs strewn between the partially burnt sumps. The soil was left with a thick covering of potash from the fires and it was into the still warm ash, that the farmer quickly broadcast his seed. Such sowing was often preceded by a piling together of scattered, half-burned logs and branches for further burning.

The seed from this first sowing of grass was highly valued. It was normal practice to sow separate patches of timothy, cocksfoot and ryegrass which were reaped with a sickle or scythe and threshed with a flail. Merchants paid good prices for such seed since a mixture of the three grasses was a sound basis for establishing permanent pastures. Men were paid about one shilling an hour for harvesting and threshing seed. Grass seed, therefore, provided a useful source of cash in hand.

Each year, in addition to the burning of logs and stumps, 20 or 30 acres of forest were felled and initially burned. A bush farmer, with about 200 acres, would be felling the last 20-30 acres on his holding after about 10 years while the remainder of the holding would be at differing stages of burning and clearing, with the original clearning being almost ready for the plough.

Ploughing was often a co-operative venture with neighbours assisting

each other or sharing in the cost of employing labour. Two men could do the job quite rapidly, one driving the horse and plough while the other attended to the roots and stumps that remained. A heavy double-furrow plough, drawn by four or five horses, was usually used. Such horses soon became accustomed to the work, learning to stop as soon as the plough hit a submerged object thus allowing it to be lifted off the obstacle. If the ploughshare got caught in a concealed stump and could not be freed, the ploughman would unhitch one of his leading horses, yoke it to the rear, and pull the plough free. For trees like rimus with large taproots, the earth would be dug out from around the stump in order to saw off the root at a depth sufficient for it to be out of reach of plough blades. In other cases holes would be drilled in particularly resilient stumps and explosives used. Team of horses with block and tackle would drag out the broken stumps.

Scrub country was brought into cultivation and pasture much more easily. The scrub was cut down with a slasher or light axe and left to be burnt after drying throughout much of the summer. The ground was

Tree felling in the Taranaki
bush, circa 1900.

ploughed immediately, although the first ploughing was sometimes fairly
rough with root masses impeding progress. However by ploughing a wide
furrow the land could be turned over quite satisfactorily. Scrubland was
inevitably very sour but application of lime soon improved fertility and
made the soil ready for grass or root crops.

Extending the Sown Grass Area: Improving the Tussock Lands

In the South Island much of the cultivable land comprised the broad,
open, tussock covered plains of Canterbury and Otago, as well as the
smaller lowland areas in Nelson, Marlborough, and Southland, while the
North Island tussock was in parts of Wairarapa and Rangitikei. The land
was relatively flat, covered in small shrubs, scrub, fern and tussock.

For many years before the introduction of the plough large areas of the
tussock grasslands in the South Island were grazed in their natural state.
Near Christchurch, Deans had described the land "... covered with
luxuriant grass ... good pasture for cattle of all descriptions ... able to be
ploughed up without any previous clearing". Elsewhere, considerable
work was needed in preparation for cultivation. Two inventions speeded

Logging, bush clearing and timber milling.

up the clearing, ploughing, and cultivation of these lands: the double-furrow plough which halved the time required for ploughing; and the portable steam traction engine and locomotive increased the efficiency of ploughing and log haulage.

The farmer aimed to get as much land as possible into good, productive, permanent sown pasture but cereals and root crops initially grew best on the newly cultivated land. In these areas, instead of putting the land immediately into grass, farmers cropped it first in wheat, oats and root crops and, as fertility began to decline, put it into pasture. On smaller farms the whole farm was sown in grass as soon as possible.

Many runholders began growing grain on a large scale, frequently making use of the contract system to break in the land. The contractor would plough the land, sow the seed and reap the harvest. However, the runholder stipulated that grass was to be the last crop sown. One of the worst features of this method was its tendency to exploit and exhaust the land. The demands of wheat depleted the land's fertility and many areas were cropped out before being sown in grass.

By 1858 Canterbury was the leading cereal producer, while during the

Above: The bush-burn landscape before logging and stumping.

Opposite page: The dramatic effect of a bush-burn taking hold.

1860s it became increasingly evident that the future of cereal production lay in the granaries of Canterbury and Otago.

A second phase of expanded cereal production began in 1875, developing into the bonanza farming era of the large estates. On some estates up to 5,000 acres of grain were harvested annually. Organising the harvest was a feat in itself and a full-time occupation for the estate owners and permanent hands. Later in the season threshing on such a scale raised problems of grain cartage and storage. Grigg of "The Levels" introduced his "train" of traction engine and wagons, while Duncan Cameron of "Springfield" had a shed constructed at the Lyndhurst railway station to store 90,000 bushels.

Many farmers had their own horse-powered threshing plants and chaff-cutting equipment but, with the invention of portable steam traction engines, most farmers preferred to have their grain threshed by contractors who moved from farm to farm. By 1880, gold production having dwindled away, New Zealand's economy rested firmly upon farm produce — wheats, oats, wool, meat, and tallow.

In the 19th century the most common grass seed mixture was ryegrass, cocksfoot and white clover, spread at about 30-40 lbs an acre. Unfortunately imperfect dressing resulted in imported seed mixtures often containing a variety of weeds and inferior seed. Included in so-called pure seed mixtures, couch grass gained a strong hold in parts of New Zealand and yarrow gained a foothold in much the same way. Regrettably

Dynamiting the tree stumps.

with such a demand for pasture and the lack of alternative mixtures inferior seed had to be used.

The responsibility for poor quality seed did not rest entirely on the British seed merchants, although faulty dressings were, in large part, responsible. The lack of sufficient seed-cleaning apparatus and short-sighted policies of some runholders in spreading cheaper seed mixtures containing many impurities, led to an alarming spread of noxious weeds — Californian thistle, ox-eyed daisy, fathen, and ragwort.

The relatively untidy appearance of so many early colonial farms was in stark contrast to their English counterparts, for with labour in the Colony being such a scarce and costly commodity, there was much less concern with weeding and suchlike. There was, however, widespread concern at the rapid spread of thistles, initially Scotch and later Californian. The hoe, bill-hook and scythe, or even judicious use of fire, were all advocated for getting rid of thistles. All the provincial assemblies passed various ordinances requiring farmers to eradicate thistles. The real problem was, however, one of enforcement, for the regulation was marked more in the breach than the observance. At one time it was even suggested that thistles

Preparations for stumping at Drummond's farm, Dannevirke, 1896.

were an excellent digestive support for livestock.

An impetus to arable agriculture came with the establishment of major agricultural machinery firms and the introduction of new and improved farm implements, ploughs, harrows, and reaping machines. Among these improved farm implements was, of course, the double-furrow plough while the importation of a new type of reaping machine, which tied the grain into sheaves with a thin wire, heralded a major revolution in harvesting methods.

Extending the Sown Grass Area: Lowlying Swamplands

The cultivation of lowlying swampy areas proved much more difficult than that of the drier, shingly tussock lands. In future dairying areas like those south-east of Christchurch and south of Lake Ellesmere, for example, much of the land comprised peaty swamp and impenetrable bog. Flax, niggerheads, toitoi and rushes grew in abundance in all swampy areas, through which it was almost impossible to move on foot or horse, while the earliest drays and wagons were able to penetrate only with the greatest difficulty. Before such swampland could be occupied and used it

After the bush-burn, seed was hand broadcast onto the still warm ash.

had to be drained and, before ploughing, it was also necessary to remove native plants too large and deeply rooted to be turned over and buried even by a swamp plough. Rubbish and debris had to be burnt and often in areas of peaty soil, the peat would catch fire and smoulder for weeks, leaving the soil very hard.

The drainage of swamps was carried out in winter while ploughing continued in the summer. A major problem in swamp clearance was the presence of submerged stumps and logs. During ploughing more timber worked its way to the surface, frequently the swing plough was jolted out of the ploughman's hands or even out of the ground as it struck a hidden log or stump. Such submerged obstacles had to be laboriously dug out and cleared and sometimes it could take 3 to 4 years before a field was finally cleared.

Eventually the swamp was ploughed, hollows and creeks were slowly filled in, drains were cut, and most obstacles to cultivation removed. After an initial ploughing in summer the land was normally left fallow until winter, ploughed once again and sown in wheat, or left until spring, when root crops were sown. With the soil abundantly rich and fertile, several crops of roots or cereals could be taken before sowing down to grass.

CHAPTER

6

Bᴜsʜ-Bᴜʀɴ

Excerpts from the diary of John Hurndall at Maungaturoto, north of Kaipara, illustrate very clearly the mixed farming practices characteristic of the bush farm. On this typical bush farm, dairying was one of many diverse farming activities contributing to subsistence for the family.

1876 January 7: cut clover. 13: made up clover. 18: made first cheese. 22: finished clover ... first stack of fodder made in settlement. 26: cut cocksfoot. 27: Arthur took four calves to Hermitage.

February 7: commenced digging potatoes, good crop. 25: went to Paparoa to committee of Agricultural Society. Arthur bought two pigs at 20/- each.

March 7: agricultural show at Paparoa, very good. 9: H Metcalfe had 8¼ cheese @ 8d. 16: Arthur burnt off at Hermitage, bad burn. 17: dressed the wheat, 6 sacks. 27: Arthur, Copley and Peter went to Hermitage to plough and log.

April 8: Arthur finished ploughing. 20: Arthur killed young bull at Worry Point. 21: settled with F Boswell for logging, 6 days @ 5/-. Arthur brought home white sow and small white boar.

May 11: F Marriner came to look at bullocks. 30: Buttercup brought in yesterday with bull calf.

June 9: killed white sow, very good, 60 lbs. 13: cows turned out. 16: turned out bullocks ... and the cows on the run. 19: J Smith killed second pig, 138 lbs.

August 8: Arthur finished fencing paddock at Hermitage. 14: Rees came to look at bullocks, ordered them all to be delivered at £12 each and 3 steers at £7/10/0. 16: ground awfully sodden. 24: made rhubarb bed, 19 roots from seed. 25: sowed pasture. 26: Arthur started to buy horses.

September 7: Tip brought in with roan bull calf. 12: Alfred Gummer brought bullocks to plough but forced soon to give up. 14: killed last pig bought off Griffin, 100 lbs. 27: sowed oats, ground in very bad order.

November 6: six lambs killed and three mangled by dogs. 8: commenced shearing. 14: finished shearing, 142 fleeces. 16: attempted to go to Waipu about wool, missed track!

December 2: Robert cut ryegrass. 7: wool shipped to Auckland. 28: E Copley shingled dairy.

1877 February 7: commenced making cheese. 28: Port Albert Show.

March 7: Maungaturoto Show. 10: finished cheese-making — seven small ones. 20: commenced potting butter.

June 6: had 12½ lbs beef from Kirk, 3½d lb. 16: Arthur found Buttercup very ill in bush. 17: the cow died from over-eating green food. 22: Arthur sent to Gummers to haul fencing. 25: Arthur sold steer for £7 ...

July 14: 28 degrees in dairy, milk frozen.

1878 January 2: Walter completed 23 hurdles (gates), 1/2d each. 4: sale of sheep and lambs, sold very low.

February 12: sold 2 steers for £6/5/0 each and 3 for £4/12/6. 14: bought 3 cows off Houldenshaw — large white, £8; light Poley, £6; brindle £6.

November 8: made first cheese, bad rennet, would not set till afternoon.

1879 January 18: £3/10/0 for 2 kauris.

February 11: agricultural show held at Matakohe.

May 8: killed 2 pigs, bought October 20 at 20/- each, sold one, 180 lbs @ 5½d lb, the other, 140 lbs, side 38, ham 18.

November 1: made 46 lbs butter and 12 lbs cheese in 8 days.

December 31: the season wet ... has produced splendid grass and crops ... stock thrive wonderfully and cows yielding a large amount of milk.

1880 January 28: Arthur went to sale of sheep at Waipu, bought 21 wethers for £9.

July 17: planted shelter pines on west side of new orchard. 20: planted new orchard, 50 apples, 12 plums, 12 peaches. 28: Bowman paid £2/10/0 for 180 lbs cheese.

1881 February 24: many burnings off and bush choppings.

August 3: Arthur commencd taking down old cowshed, built 17 years ago.

September 2: first cattle sale day.

1882 June 6: bacon and hams sent to George. 14: commenced fowlhouse, pig pen and pig sty.

September 15: sold 2 steers and 3 heifers.

1883 December 8: Arthur bought 66 ewes and wether hoggets, 6/-, and 2 pigs, 10/-.

1884 August 1: Arthur bought 2 white pigs, 18/-. 6: ... bought 2 cows and 1 calf, £7/10/0 each. 7: exchanged 2 yearling heifers and £1 cash each for Berkshire sow.

The diary entries illustrate the extremely diverse nature of the bush farm economy as well as the high degree of seasonability in farm production patterns. Records of the early Maungaturoto settlement for 1 January 1869 reveal many similarly involved settlers.
Maungaturoto was extremely isolated. Some of the cattle originally came from settlements to the west in the northern Wairoa district, and in the

Dairy farming, circa 1900.

east from two farms at Kaiwaka which had a relatively large area in grass and possessed a large herd, mainly shorthorns, with a reputation for high quality. Cows could not be driven to Maungaturoto but were roped and led one by one.

For several years dairy produce brought high prices but there was only a limited local market for butter and cheese. Those fortunate enough to own cows supplied themselves with milk and dairy products, while those without bought their neighbours' surplus. Prices were high but demand was limited; any surplus butter was sent to Paparoa.

In 1867, there is a first reference to the possible formation of a local agricultural society for Maungaturoto and Paparoa. At this time there was a population of just over 100, with 172 cattle, 63 pigs, 8 horses, 175 acres under sown grass, and 24 acres in wheat.

Taranaki in the 1880s and 1890s

Roberts Wells' Taranaki bush farm was inland from Waitara, about 10 miles from New Plymouth, and was developed on two 50-acre sections

A "Taranaki gate" or a sub standard fence?

Post and rail fencing was common in many areas.

allocated in 1874 in return for military service. The first task was to subsist while beginning early bush clearing. As on most bush farms in the North Island during the early years, some supplementary income in winter came from road making or bush felling on contract, or from cutting roof shingles for sale. In summer farm work was done for neighbouring settlers on a labour exchange basis.

By the 1880s farm work was becoming more important. Most of the bush had been burnt or felled and preparation was underway for ploughing, cropping and finally, the sowing of long-term pasture. Fencing the paddocks with post and rail involved splitting the posts and rails out of tall timber in the bush, mortising the posts, pointing the rails, and transporting them to the fence site for erection.

Slab and shingle dwelling amid
the remains of bush-burn.

The pioneer bush farmer in the 1870s and 1880s had to remain solvent
while carving out his farm from the bush, but the amount of money that
changed hands was minimal. In 1876 Wells' largest receipt was from the
sale of the main crop, 1,240 lbs of grass seed which sold for 6d a pound.
Sale of potatoes was another source of income. Kegs of farmhouse butter
were bartered for necessary provisions. General wages averaged 8/- a day
but could occasionally rise to 15/-, while loans were almost impossible to
obtain. Early trading was based primarily on bartering with merchants. As
late as 1882 Wells was still dependent on barter, for instance 6 cwt of
potatoes in return for repairs done to a cart axle, while butter continued to
be credited against outstanding accounts at the local store. An unusual
source of cash was the gathering of fungus. This was collected from
stumps and logs and sold to Chew Chong who paid 4d a pound and 5%
commission.

By 1889 butter was no longer bartered — an important advance.

The Catlins, 1892.

However, butter was difficult to sell in New Plymouth during the seasonal flush, even at very low prices. Attempts were made to open up markets elsewhere. In March 1889, Wells sent six 60 lb kegs of salted butter to Sydney for a gross return of 1/- a pound compared to 8d to 8½d received in New Plymouth. Another consignment, sent to the Wanganui area, returned only 8d a pound. As these markets were liable to considerable fluctuations, the trade was highly speculative. Fresh butter was in demand during winter, but by October the price had fallen to only 6½d a pound. The market demand was for salted butter in kegs, usually sold to traders who advanced 6d a pound on it before sending the produce to England. By December the cash advanced on butter for the English market was reduced to 3d a pound and, with the need for cash, kegged butter was sold on the New Plymouth market for 6d or 7d.

At the beginning of the 1890s the margin between financial success and failure was very finely drawn. Despite this, it was in this decade that Wells' bush farm was transformed from a bare and struggling subsistence farm, with mixed crops and stock, to the beginnings of a profitable commercial enterprise and an all pasture dairy farm, with cows, fences and fodder crops.

By 1900 the pioneering phase had ended and farming had entered a new and prosperous era. Butter was no longer laboriously churned in the home dairy for sale on uncertain markets at marginal prices. There was now a regular monthly milk cheque from the local dairy factory and an annual bonus paid out in winter when the milk return was smallest. Milk was now the major source of income.

Concentration on Dairying

The 1890s saw the development of an all pasture farm. By 1900 the temporary pasture association of cocksfoot and ryegrass had been improved by the addition of timothy, cowgrass, and clover, a seed mixture which was sown with various fertilisers. Only towards the end of the decade was fertiliser applied, by which time Wells was well on the way towards an all grassland farming economy.

By 1889 the quantity of milk produced was so great that an entire morning each week was often needed to churn it, salt it, and put it into kegs ready for the market. Sometimes, during the warmer weather, butter had to be put down in the well to cool before it could be fully worked. On Wells' farm dairy cows grew in number from 16 to 24 and increased the average annual production of butter per cow from 142 lbs in 1889 to 200 lbs in 1900. The principal reason for this increase was that the home farm dairy was supplanted by a local factory which, despite numerous mechanical faults and a small milk supply, was a great improvement for dairy farmers. By the end of the decade the increase in stock numbers was providing surplus young stock for sale.

Otago Peninsula 1865-1871

Walter Riddell was an early bush farming pioneer on the Otago Peninsula. The diary account of his farm life and work, 1865-1871, based upon day to day activities, provides a useful picture of one of the Highcliff "co-operative eight". Riddell arrived from Scotland in 1863 to buy 90 acres of bush covered land at Sandymount on the Otago Peninsula. Setting out from Dunedin in March 1865, he and a carter took "... ten days to carry our things through the bush ... to the home, comprising one room, 14ft by 12ft, built of fern tree" — a distance of perhaps no more than 6-7 miles. Two weeks later a ton of potatoes took 4 days to be transported through the bush. His earliest work clearing and burning the bush and scrub on his own peninsula property as well as contract clearing and felling for neighbours.

Early in 1866 he put up a bail for a cow, bought the previous November for £12, and which later had a heifer calf in January. In February he sold 4 lbs of butter to a Dunedin grocer for 1/8d a pound and a dozen eggs at 2/3d a dozen. In March of that year he built a pigsty, bought a pig for 16/-, and sold 41 bushels of grass seed for 7/- a bushel. Later in March he brought back the heifer from Green Island, completed some fencing, and put the cow out to graze nearby, paying £2.8.0 for 24 weeks' grass.

In January 1867 he records burning bush, including a "good burn" which destroyed most of his lower fence and almost burnt the byre, while in April, as an odd-job carpenter, he begins to build a milk house and byre for another settler; during November he made several butter churns for other farmers nearby.

Early in 1868 he was not only busy cutting and preparing timber for a local church he was to build, but also cleared bush, planted crops, made more butter churns, and built some small cow byres. During March he sold a bull calf for £2.5.0, and lent a neighbour £8 to help him buy a cow.

In 1869 he spent much of the year clearing bush and cutting timber, digging potatoes, sowing turnips broadcast, cutting hay, raking grass seed, and making churns. By January he wrote that he now had eight cows and later in the same year attended a meeting organised by John Mathieson, a neighbouring farmer, taking three shares to become one of the eight shareholders in the first dairy cheese-making co-operative.

Riddell's diary indicates the variety of work associated with early bush clearing to develop a farm, the extent to which early settlers assisted one another, frequently working for neighbours for additional cash income, and overall, the relatively small scale and non-specialised nature of early bush farming which provided a large measure of subsistence.

The British Agricultural Model

A particular feature of early farming development was the continuing conflict of ideas as to which were the most suitable kinds of farming practices for the special New Zealand conditions. This was evidenced in the frequent reports and letters in such papers as the *New Zealand Country Journal* published in Christchurch. Many settlers, as well as agriculturalists and journalists, believed the real future for New Zealand agriculture lay in creating something identical to an English mixed farm, complete with a wide range of crops and livestock, and accordingly were highly critical of the "primitive" colonial farming practices often adopted. A number of these conflicting notions on the kinds of farming suitable for New Zealand in the 19th century have been recorded.

Wakefield was one of the first to advocate the English style of "clean cultivation, labour intensive" agriculture, organised through so-called capitalist landowners, as an integral element of his scheme of systematic colonisation. However, he failed to appreciate the scarcity of agricultural labour and many of the "capitalists" were absentee landowners reluctant to settle in the Colony.

Throughout the early period, sown grass continued to be the principal crop on New Zealand farms. The total area of sown grasses increased by 142,400 acres in the period 1851-1861, by 618,000 acres from 1861-1871, and by a massive 2,782,900 acres in the decade, 1871-1881. The most notable expansion of sown grass occurred in the Tamaki isthmus, in coastal Manawatu, the Hutt Valley, the area south of Napier, the Wairarapa in the North Island, and in the eastern coastal lands of the South Island, all of which had formerly been covered in scrub, bush or fern. By 1881 about 60% of all sown pasture in New Zealand had originated from surface sowing into uncultivated bush ashes.

The merits (or otherwise) of surface sown pasture provoked bitter conflict and contention for, like many other farming practices in 19th century New Zealand, the notion did not conform to the traditional English method of clean cultivation. Initially, it was widely criticised as being unsatisfactory compared to the English practices of careful preparation and cultivation prior to sowing. Many overseas observers, and indeed commentators in New Zealand itself, censured the colonial practice of direct surface sowing on to the still warm ashes after bush burning. They failed to appreciate that the latter system had evolved under conditions quite different from those with which they were familiar. One writer in the *New Zealand Country Journal* in 1881 referred to New Zealand's sown pastures as " . . . a disgrace to the country, or rather to the farmers, for the fault lies with them". Seemingly with the tidy English hedgerows and green fields in mind, the author continued that all farmers "should harrow, roll and manure their grassland regularly" as in England. Writers such as this failed to recognise the need to adapt, and if necessary modify, the traditional methods when confronted by the totally different environment of the young Colony. Furthermore, labour was scarce and even when it was available was too expensive for the method of clean cultivation to be practicable.

These conflicting ideas on the need for proper cultivation as a precondition for effective sowing of grass seed must have also contributed to the conflict in Taranaki in the 1850s. This clash of principles was implicit in the contrasting evaluations made of bush on the one hand, and fern or scrub on the other, although these differences were due in part to the cost of the labour input required for the clearance and cultivation of the different types of land.

In the 1850s and early 1860s there was also a common tendency to sow

Typical outdoor milking scene.

pastures with a single variety of grass seed. However by the later 1860s and 1870s the sowing of grass seed mixtures had become widespread practice throughout the Colony, with each district or area, indeed almost every individual farm and farmer, preferring a particular and specific mixture. It is recorded around Auckland, for example, that 2 bushels of grass seed per acre would normally include 1 lb perennial ryegrass, 4 lb red clover, 6 lb white clover, as well as various quantities of cocksfoot and peas as available.

Another misguided notion shared by many of the early settlers was that cattle needed supplementary winter feed as in England. Hay was widely used, for it was relatively easy to prepare, needing only a fairly low labour input. However, the total hay acreage had grown from less than 30,000 acres in 1851 to only 54,000 acres in 1881. Consequently the *New Zealand Country Journal* reported that too many colonial farmers were neglecting a vital part of the farm economy, namely hay, in spite of the ready and wide availability of mowing machines by the late 1870s. Such commentaries suggested an almost paranoic obsession with English farming practices as well as a total failure to recognise the local emphasis upon supplementary feed crops such as turnips and rape.

Early bush farm stocking about 15 cows, a few calves and some hens.

The modern concept of all grass farming, in terms of dairying and fat-lamb farming, was viewed with a great deal of scepticism in most 19th century farming circles. There was not only the real problem of fluctuating dairy and livestock prices (mixed farming spread such risks over a wider range of products), but the internal economy of the farm was upset, for the absence of arable cropping, especially grain, meant there was no straw for stalling or root crops for winter feed. These arguments again suggested a fixation with English farming methods and a consequent failure to recognise the needs and merits of alternative practices. Straw, of course, was seldom used and rarely saved since the stock were never stalled or housed and thatch was never needed. Similarly, at least until the late 1870s, any North Island farmers growing winter feed, apart from grass for hay, were in a clear minority because abundant feed was readily available in the nearby bush and fern.

Conflicting ideas also arose over milking practices. In the Colony cows were put into bails, a rear leg was roped across to safeguard the milker from being kicked and to prevent the cow from overturning the milk pail. This method of securing cows, although almost universal in the Antipodes, was scarcely known in Britain where dairy cows were regarded as highly sensitive to even slight changes in milking routines.

Among many settler-farmers there was a growing recognition that

Bradfield homestead, the Catlins.

pasture land and grass paddocks which were continually grazed and occasionally topdressed lasted longer and were not quickly exhausted by constant grazing by livestock. As such, this practice proved more economic since in the long-term there was less need for resowing the pasture. The traditional dung and straw of the English farmyard was, however, not readily available because of the absence of stall feeding and winter housing of livestock, and of course, free ranging stock posed unique problems for manure collection.

Therefore, New Zealand farmers had to consider the possibilities of manures such as guano, referred to in a local newspaper as far back as 1877 as the ''most popular'' on New Zealand farms. As was to be expected, critics who loudly condemned colonial farming practices attributed many of the failures to eschewing the English method of dependence upon farmyard manures.

Overseas commentators were also widely critical of the failure of New Zealand farmers to adopt a proper crop rotation system. Whether referring to North Island bush farms or South Island grain farms, such critics were quick to note that in such circumstances it was no wonder productivity was declining, either in terms of yields or stock carrying capacity. In their view the only proper and long-lasting solution lay in the adoption of an appropriate English style rotation system. In reply it was pointed out that such a solution was unnecessary on pioneering colonial farms since local market demand was not so exhaustive on the land as in England.

7

REFRIGERATION

Refrigeration was decisive in committing New Zealand to make a major shift in the overall direction of its economic growth. The possibility of exporting perishable goods provided economic viability for new forms of pastoral farming, namely dairying and fat lamb raising. Before 1882 the range of New Zealand's main exports had been severely restricted by the requirement that they be non-perishable in order to withstand the lengthy sea voyages. Furthermore, before the 1880s the expansion of dairying and dairy processing had been constrained by the limited scope offered by the domestic market. If the factory system of dairy production was to be adopted it required the assured demand of an expanded market, a condition made possible only by the development of refrigeration.

Introduction

Until the 19th century interest in refrigeration as a method of preserving perishable foodstuffs had been sporadic. From 1819 onwards, several patents for refrigerating processes were recorded in England, until by 1853 there were about 10 separate processes known, although none was in use on a commercial basis. At that time, Britain's urban and industrial population created an ever increasing demand for a regular meat supply and this need could be met neither locally nor by importing livestock. By contrast, several countries in the Southern Hemisphere had large surpluses of livestock which normally had to be disposed of by the wasteful process of boiling down for tallow. In the absence of any effective form of preserving food for the long sea voyage, perishable export cargoes such as meat were quite impracticable.

One practical solution emerged in the form of canning meat and this was especially successful in Australia. In 1869 the United Kingdom imported 2 million lbs of tinned meat, mainly boiled mutton and corned beef, and by 1880 imports had risen to 16 million lbs. New Zealand's ventures in canned meat exports did not prove highly successful. However, the real need continued to be for fresh meat, and it was to preserve meat during the lengthy sea voyage that led to refrigerating processes being researched and developed. Alternative processes were also investigated. A shipment from New Zealand of legs of mutton packed in tallow arrived in Britain in good condition but proved to be commercially unsuccessful. Of all the approaches, freezing seemed to offer the best chances of long-term success.

Milk collection by ferry.

The feasibility of freezing was clearly demonstrated by the work of two pioneers in Australia, Thomas Mort and James Harrison, the latter seemingly years ahead in developing and patenting a practicable refrigerating process. In 1873 Harrison publicly demonstrated his freezing machine in Sydney, freezing carcasses of beef and mutton, as well as fish and poultry, to be consumed at a banquet 6 months later. Later in the year the refrigerating machine was fitted on the S S *Norfolk* to send a cargo of frozen meat to London but the meat arrived in unsaleable condition.

Thomas Mort patented his ammonia absorption refrigerating process in 1867, 6 years after he had established Australia's first freezing works at Sydney. In 1875 he declared Australia was destined to become Europe's major food supplier, but a year later when he chartered and equipped the sailing ship *Northram* for a trial shipment, the machinery failed before the vessel sailed and the cargo of meat was unloaded. Harrison and Mort had both demonstrated the possibility of freezing meat to preserve and export it, but neither had successfully transported a shipment.

Other refrigerating experiments were continuing elsewhere. A machine

Okains Bay wharf.

for freezing meat invented by Carre in 1860 was installed on the S S *Paraguay* in 1877 for a shipment from Buenos Aires to Le Havre. This shipment arrived in 1878 in perfect condition and was the first successful effort to export frozen meat. A further inventor, Tellier, concentrated upon chilling rather than freezing meat and in 1877, the *Frigorifique*, fitted out with chilling machinery, successfully shipped a consignment of chilled meat from Buenos Aires to Rouen in France.

The Bell Brothers worked on alternative refrigerating processes and in 1877, took out patents for the Bell-Coleman Refrigerating Company. Their machinery was fitted on many ships engaged in the early refrigerated trade from New Zealand. The Bell-Coleman refrigerating plant was also fitted on the *Strathleven* when, in 1879, some Queensland squatters, impressed by the success of the earlier *Paraguay* shipment, loaded on board 40 tons of beef which arrived in good condition in London in 1880.

Refrigeration and the early New Zealand Export Trade

Among those who inspected the successful *Strathleven* shipment upon its arrival in London was the producer-representative of the New Zealand Loan and Mercantile Agency Company. Encouraged by a favourable report to the Company on the condition of the meat, William Soltau Davidson, head of the New Zealand and Australia Land Company, decided to investigate the possibilities of the frozen meat trade for New Zealand. Like all other companies in the country Davidson's company was ridding itself of thousands of surplus livestock at ridiculously low prices every year. The stakes were high indeed when the company undertook to find a cargo of 6,500 ship carcasses and the shipping company Shaw Savill & Albion fitted up the 1,200 ton sailing vessel SS *Dunedin* at no small cost.

The first successful shipment of frozen meat left Port Chalmers for London on 15 February 1882. The cargo comprised 3,521 sheep and 449 lamb carcasses from the Company estates, and 939 sheep and 22 pigs

supplied by others. This first shipment not only opened the way for the meat trade of the future, but also laid the basis of a dairy export trade following the success of the first small shipment of butter to London. The butter for the experimental consignment, produced in the Edendale dairy factory set up by the New Zealand and Australia Land Company, arrived in the United Kingdom in good condition. In spite of sundry mechanical mishaps and delays in loading, as well as an extremely long voyage of 98 days, it arrived at the London docks on 24 May.

The people most prominent in this first butter export were WS Davidson of Edinburgh, General Manager of the New Zealand and Australia Land Company's Scottish enterprises, and Thomas Brydone, the New Zealand based Superintendent of the company. Brydone had been responsible for most of the organisational work associated with the New Zealand side of the venture and it was due to his influence that a small trial shipment of butter was included with the cargo of frozen meat.

The second consignment of butter shipped under mechanical refrigeration left in the *Lady Jocelyn* in 1883, when seven kegs of export butter were included with the first cargo of frozen meat sent from Wellington. From these small beginnings the proportion of the total value of New Zealand exports accounted for by butter and cheese rose from 0.17% in 1880 to 7.33% in 1900.

Refrigeration and Overseas Shipping

The possibilities of setting up a steamship service between New Zealand and Britain had been discussed before the first successful shipments of frozen meat and butter in 1882. In 1881 the New Zealand Government had offered a subsidy of £20,000 annually for the establishment of a regular steamship service to England, a voyage of up to 50 days; the

incentive, however, proved too slight to attract any shipping owners.

After the first successes, many "modern" vessels were soon fitted with the new refrigerating machinery, while several steamers were specially designed and fitted for the purpose of transporting frozen cargoes. The New Zealand Shipping Company had five steamers built in the mid 1880s to do the "Home" run in 45 days out and 42 days return. Such steamers were fitted with refrigerating machinery and cool chambers, and were capable of carrying 12,000-13,000 carcasses of mutton and a comparable quantity of dairy produce.

In 1882 the New Zealand Shipping Company fitted up its first steamer the *Fonstanton*, with refrigerating machinery, in 1883 it began its first regular steamship service with the *British King*, the latter fitted with Haslam's cold air refrigerating machinery. Later in the same year the *British Queen* inaugurated a monthly service. The steamships, *Ionic, Doric* and *Catalonia* were chartered in 1883, and were indicative of the optimism in the future of the refrigerated trade. By 1884 the Company had its own 4,500 ton refrigerated steamers *Tongariro, Aorangi, Ruapehu, Kaikoura* and *Rimutaka*.

The Shaw Savill and Albion Company, involved in the New Zealand trade since the 1850s, had shared in the inauguration of the frozen meat trade in the *Dunedin* in 1882. In 1883 the Company ordered the construction of three 5,000 ton steamers, specifically equipped for carrying frozen meat and other perishable cargoes. Together, the New Zealand Shipping and Shaw Savill and Albion companies maintained a regular, fortnightly steamship service between New Zealand and London.

Other shipping companies also entered the refrigerated trade — the Tyser and Star Lines with 14 steamers by 1900, and the Federal Line with a fleet of 10 steamers by 1903. In addition to fleets of new steamers, there were also many early sailing ships re-equipped with refrigerating machinery. The latter took 100-110 days for the voyage compared with 40-45 days by steamer.

Such shipping fleets provided regular overseas shipping services, but there were problems associated with the varied conditions under which meat and other perishable produce was carried, and with freight charges. During the 1890s there were increasing numbers of damaged cargoes, partly due to deficiencies in early refrigerating equipment, although the shipping companies gradually installed more up-to-date and improved machinery. However, probably most critical of all was the fact that, while in early vessels freezing was carried out on board, in later steamers, the cargo was frozen nearby on shore. This meant it was frequently insufficiently frozen before storage in the holds, thus arriving in the United Kingdom in bad condition.

The freight charge on the *Dunedin* shipment was 2½d a pound, including freezing charges. However, with the growth in refrigerated trade and the introduction of steamers with vastly improved machinery, storage and increased refrigerating capacity, freight rates began to fall.

The development of refrigerated shipping fleets ensured that perishable dairy produce could withstand the voyage and arrive in excellent condition for sale and consumption in British markets. But this presupposed that it had arrived at the port of export in comparable condition. Such quality could only be assured when appropriate refrigeration and cool chamber facilities were extended to coastal shipping and railway services, as well as factory and port coolstore facilities. This guaranteed an efficient and continual refrigerated transfer from factory to port and thence to market.

Opposite page above:
Port Chalmers.

Opposite page below:
Wanganui Wharf, 1906.

Taranaki Co-op cool stores.

Refrigeration, Coastal Shipping and Inland Water Transport

Coastal shipping was a significant problem for the early dairy industry. Dairy factories were not always located near ports, nor did they always have direct rail links to such ports. Instead, they had to be located as near as possible to their source of raw material, milk. Dairying gained special prominence in Taranaki at a time when coastal shipping was the only link with Wellington. Later when rail links were established to tie in with the main North Island network, coastal shipping links through ports such as Patea were still used for exports of dairy produce because they were cheaper.

Dairy development in the Waikato area also predated major railway development and rivers played an important role in early transport, the Waikato and Piako rivers being significant factors in the siting of some early factories. Water transport was also important in both the Kaipara and Bay of Islands. Before the extension of the railways into North Auckland coastal ports as well as inland ports like those on the Kaipara played a major part in the transport of milk and dairy produce.

During the early years of the dairy export industry Wellington served as the main shipping port for exports overseas. This meant that in the absence of an integrated railway network, all dairy produce from the provincial dairying districts was transported in small coastal vessels to the port of Wellington for transfer on to overseas vessels.

Unfortunately there was a marked delay in the spread of refrigerated handling and storage facilities from ocean to coastal shipping and, consequently, some years elapsed before coastal vessels were fitted with insulated chambers and storage, let alone refrigerating machinery. In the meantime, dairy produce travelled as ordinary cargo. A report of 1896 referred specifically to the difficulties of shipping export dairy produce on coastal services.

Refrigeration and Railways

The development of many remote areas was brought about by continuing railway growth and expansion. Although early dairy development in the Waikato district was based more upon river transport the growth of rail transport was critical over much of the area and its continued development was a condition of future progress since it was only then that the farmers' reliance on coastal shipping was reduced.

Refrigeration put new demands upon railways. The developing meat industry required refrigerated trucks and wagons to carry frozen carcasses from the freezing works and coolstores to points of shipment in the larger ports. The dairy industry eventually demanded tarpaulin covered trucks and later, distinctive individual cool trucks. All industry naturally sought transport that was as cheap as possible while proving the requisite standard of service. The recognition that the Railways Department gave to these expectations, signalled the increasing realisation that the potential growth of the primary industries was important both as a source of revenue and as a critical factor in the development of the Railways themselves.

Collection of milk on the Waikato River for the New Zealand Co-operative Dairy Co.

The specific needs of the frozen meat trade were promptly met with the provision of 33 special meat trucks by March 1883. By March 1886 a further 53 refrigerated wagons were reported. The early demand for refrigerated wagons was slow and in a 10 year period the total number had risen to only 65. However, the rapid development of dairying at the turn of the century saw an equally rapid demand for specialised railway rolling stock. Soon there were 208 special refrigerated rail truck wagons in use, increasing to 426 by 1918.

Although the Railways Department met the specialised needs of the frozen meat trade fairly quickly, the slow growth of the dairy industry made it less responsive to their demands. In 1891 the Department was urged to provide insulated cool chambers in trucks for transporting dairy produce to the ports. By 1892 the Railways Department had 113 ventilated cool trucks.

The need for improved, more efficient and more effective railway

The "new face" of dairy shipping.

facilities for dairy produce continued and from 1905 onwards there were improvements. In south Taranaki, however, dissatisfaction with the existing transport facilities and service resulted in the local dairy companies joining forces to form their own shipping company. This took over some of the insulated and refrigerated vessels then carrying dairy produce from Patea to Wellington for shipment overseas. Additional vessels were acquired and the company was soon handling all the dairy export cargoes from South Taranaki.

In 1896, the Chief Dairy Expert reported that the experiment by the Railways Department to provide iced trucks had proved to be "fairly successful" but there was still room for further improvement. By 1898 the Department was being praised for "... providing a regular service of ice vans three times a week ... for all through the season butter has arrived at coolstores in most satisfactory condition when forwarded using these ice-cars, the larger portion being firm and cool". By 1904 dairy growth and expansion was creating a growing demand for improved rail facilities, and after this date the cool truck services of the Railways Department seem to have satisfied members of the dairy industry.

The importance of the early railway network for moving both cream and dairy produce was reflected in the location of many dairy factories. In the South Island the skimming station activities and networks of the Taieri and Peninsula Milk Supply and Canterbury Central Co-operative Dairy Companies followed the existing rail network very closely, while in both islands many factories were located alongside or close to railway sidings.

8

DAIRY FACTORIES

In the late 1870s and early 1880s, many were agreed on the need to adopt a factory system of dairy processing but the limited size and scope of the domestic market scarcely provided the necessary incentive for such a move. Dairymen, government officials, industry leaders and newspaper editors had all expounded the obvious advantages offered by the system. In 1881 the government of the day offered a bonus of £500 for the first successful export of factory manufactured butter or cheese, a real and attractive inducement. It was the development of refrigerated shipping that proved decisive, it opened up the full market potential of the British market and paved the way for the introduction of factory processing.

The large-scale requirements of the British market could not be met by the traditional approaches to butter and cheese manufacture. In terms of both quantity and quality traditional methods were inadequate and a profound change was required. Only the factory system could ensure an increasing supply and a regular volume of dairy produce, along with the quality control necessary at every stage of production.

The Factory System of Dairy Production

Apart from such elements as larger-scale and more mechanised processing methods, the essential characteristic of the factory system was the collection of whole milk from several nearby suppliers for processing in a central location, normally specially and better equipped than the farmhouse dairy. Here the special skills and expertise of experienced dairymen could be utilised more fully and shared for the mutual benefit of all the suppliers involved by virtue of the increased demand and higher prices realised for improved and reliable quality.

In New Zealand, beginning in the 1840s, Banks Peninsula cheese exports had essentially been the product of larger, improved farmhouse dairies. One of the first recorded instances of any form of "factory" processing, probably closely akin to the localised early "cheese circles" of Wisconsin, is the reference to Taranaki in the early 1850s where milk was collected from several neighbouring farm suppliers for commercial cheese-making for local sale.

The dairy factory system, as it developed after 1882 was not a new phenomenon in international terms. Some of the earliest developments, in the special kind of factory system that was introduced into New Zealand, took place earlier in the United States and Canada. In 1851 Jesse

Top of page:
**Falconer Cheese Factory,
Collingwood.**

Above: Highcliff circa 1890.

Williams, the founding father of the American cheese factory system had built his first factory at Rome in Oneida County in the state of New York. Whole milk was collected from surrounding farm dairies for processing and converting into cheese at a central location. Factory processing meant improved quality, increased demand, higher prices, better use of plant and appliances, the economic advantages of large-scale operations and improved conditions of handling milk.

THE RIVERDALE DAIRY COMPANY, INAHA, TARANAKI.
LARGEST CHEESE FACTORY UNDER ONE ROOF IN THE WORLD.

In 1866, the leader of the American Dairymen's Association visited Scotland and England. His main purpose was to view progress in cheese-making, particularly the work of Joseph Harding and his family in Somerset who were attempting to systematise a process of making Cheddar cheese. He recalled that in the United States, at that time, there were 2,000 small cheese factories receiving the milk supply of more than 200,000 dairy cows.

Riverdale Dairy Co.

One notable difference in the early growth of factory dairying in New Zealand and Wisconsin, lay in the contrasting roles of cheese and butter manufacture. Wisconsin factory processing remained almost exclusively concerned with cheese-making. Butter-making remained a predominantly farmhouse based operation; it was easily made by most milk producers but cheese-making, even on farms, called for more experience and expertise, it was thus more suited for factory production.

The situation was quite different in New Zealand. Although the Highcliff factory, established in 1871, was a reasonably successful enterprise, most of the less successful attempts to establish factories had revolved around cheese-making. In the 1880s the majority of the earlier factories were equipped with dual plant to enable production of either butter or cheese. Dual plant factories were more flexible in operation and so were more easily able to adapt to meet new supply and demand

Ngaere Dairy Factory circa 1900.

patterns. The first dual plant factory began operating in 1882. There were three of these in 1882, a further nine in 1883, and another eight in 1884, the year in which the first dairy factory, equipped exclusively for butter-making, was established at Karere at Longburn near Palmerston North.

In examining early factory adoption, two fairly widespread transitional phases may be noted as they were significant in phasing in the new rural production system. While farmhouse dairying was still dominant in most areas, an early development saw the frequent establishment of many small packing stations, to process small quantities of farm butter surplus to local or family needs. In the packing stations the small amounts of frequently granulated farm butter that were available were blended for marketing in nearby towns. The system survived as long as farmhouse butter-making persisted with packing stations normally located in or close by larger settlements.

Well designed and fully equipped dairy factories did not appear immediately, often there was a transitional phase during which a new centrally located structure was erected to process milk collected from several suppliers, but the processing methods used were still the traditional farmhouse practices. Similarly, the equipment and appliances of the old farm dairy were used until they could be replaced by more up-to-date plant. Larger, steam-driven equipment such as mechanical separators, had of necessity to be set up in new factories rather than smaller dairies.

The changeover from farmhouse to factory did not immediately take

Awhitu, Hamilton Brothers cheese factory, circa 1899.

place nationwide, nor were all the farmers equally convinced of the benefits of the innovation. Thus, in the early years, it was not unusual to find farm and factory processing surviving side by side. In other situations the physical terrain created special difficulties for the ready transport and supply of milk for processing. In such districts, where roads and tracks were not available even for wheel-less sledges, or in very isolated districts far away from factories, the milk was retained on the farm for processing long after factory processing had been adopted elsewhere. Thus farm-house butter and cheese manufacture persisted in the more rugged and isolated districts well into the 20th century.

Another point worthy of note in founding the earliest dairy factories was the decisive role played by both individual and corporate entrepreneurs in an industry long renowned for the overwhelmingly co-operative nature of so much of its enterprise. Without such initiative and enterprise, and a readiness to invest and risk their own capital, the economic viability of the dairy factory system would have been much delayed. Highcliff excepted, the earliest efforts at dairy co-operation had one element in common, a singular lack of success. This was probably due to a lack of business acumen and administrative experience as well as inadequate technical expertise. More particularly it was also due to limited capital resources. It was therefore left to proprietary interests to initiate many new dairy developments. For several decades the dairy proprietors played a leading and very decisive role in the industry, although for a long time their activities were much under-rated and their contribution seldom acknowledged. In the 1880s and 1890s dairy

co-operatives grew only slowly in number compared with proprietary interests and did not even constitute a simple majority of the factories operating until the turn of the century.

However, as quite a number of dairy proprietors ultimately found out to their cost, once the economic viability of factory dairy processing was proven and established, farmer-supplier dairy co-operatives could scarcely wait to take over control of the operation, often with scant regard for the proprietary interests that had made the venture possible.

The Earliest Dairy Factories

Apart from the early cheese-circle type of factory venture in Taranaki in the 1850s, the first documented effort at factory cheese-making, as well as the first real initiative in dairy co-operation, was in 1871 at Highcliff on the Otago Peninsula. On 22 August of that year, John Mathieson called a meeting of eight other local settlers who owned dairy cows, to discuss the formation of a dairy company for the purpose of "cheese-making on the co-operative principle". Those attending the meeting took up shares to ensure a guaranteed supply. The first suppliers duly delivered milk to the new factory, a wooden outbuilding on Mathieson's farm, and dairy operations commenced in September 1871. In 1875 operations were re-located to a new building at Highcliff. Both the cheese-maker and his wife were experienced in, and had the equipment for making, the

Above: Stanley Road dairy factory, Stratford.

Opposite page above: One hundred pounds of butter being worked in the Hautapa factory.

Opposite page below: The first Radiator Dairy Factory in New Zealand.

77

Part of the original barn and byre complex at "Springfields".

traditional Scottish "Dunlop" cheese. The factory continued to operate as a cheese factory for some years, known variously as the Otago Peninsula Co-operative Cheese Factory Company Limited (Minute Book cover), Peninsula Cheese-making Company, Pioneer Cheese Company and Peninsula Pioneer Cheese Company, until in about 1884-85 it was converted to processing granulated butter from local suppliers.

During the 1870s other efforts to set up small cheese factories were briefly recorded but few, if any, details are available. They were both co-operatively and privately operated but none survived and functioned as long or as successfully as Highcliff.

9

DAIRY PROPRIETORS

The early and widespread responses to the demand for factory dairying owed much to the actions of various proprietary interests. Such proprietors demonstrated their faith in the future of dairying by investing their own capital in factory construction and in the introduction of factory dairying. Without such initiatives the growth and development of the dairy industry might have been seriously delayed. The depressed economic conditions of the 1880s were scarcely conducive to farmer initiative and dairying investment. In fact farmer groups became interested in taking over many proprietary dairy factories as co-operatively owned and operated enterprises only after the original private initiatives had proved the value of the new system of dairying.

In the late 1870s and early 1880s there had been isolated efforts to set up dairy factories on a co-operative basis as a means of fostering the factory system, but with the notable exception of Highcliff the ventures had failed. In many instances such failed dairy ventures were taken over, reorganised and profitably operated by dairy proprietaries under private management and ownership.

In trying to assess the significance of proprietary initiatives and investments, it is important to bear in mind the comparable contributions of co-operative factories and their share of production. After the financial failures and closures of many early dairy co-operatives it was possible for the Chief Dairy Expert, in his report for the 1892-93 season, however, to note that in that season "... every one of our co-operative factories is now in a healthy, financial and operational state." The co-operative movement grew steadily as proprietary interests slowly declined in importance until in 1894 nearly 39% of dairy factories were co-operatively owned and operated.

Towards the end of the 1890s there was a definite leaning towards dairy co-operation and proprietary factories were gradually taken over by suppliers to be organised as co-operative companies. The transition was either carried out most amicably or with bitterness and resentment, on the part of both parties involved.

The majority of all dairy factories were co-operative for the first time only in 1903, cheese factories slightly earlier in 1899 and creameries in 1904. By 1906 more than 60% of dairy factories were co-operative and over 80% by 1917 as the move to co-operation progressed steadily. The industry was totally co-operatively-owned and operated as recently as 1951.

Defiance dried milk factory at Bunnythorpe.

An interesting but very small-scale development in the 1980s has seen the re-emergence of private interests in several small farm-size dairy enterprises like "Settlers" dairy produce in Barrys Bay. Most of these ventures are concerned with producing more highly-specialised cheeses.

Proprietary dairy interests were of two kinds, individual and corporate. Individual dairy proprietors came from a variety of backgrounds, although many had had their introduction to the butter business as storekeepers in the butter barter trade, buying farmhouse butter for blending. As individuals, however, some had the necessary personal capital means, or at least the associations, to underwrite the spread and development beyond a single factory. As the butter industry grew with its rival alternative creamery and farmhouse separation production systems, the attitudes of individual proprietors and their own personal preferences became largely responsible for patterns of dairy developments in some areas.

Apart from individual dairy proprietors, a small number of corporate proprietors also played extremely important roles in various dairy initiatives. The New Zealand and Australia Land Company, for example, set up the pioneer prototype dairy factory at Edendale; Nathans of Bunnythorpe pioneered milk powder; while at a time when most dairy factories refused to accept farm-separated cream for butter manufacture, the Wellington agent of one farmhouse separator set up its own factory to handle such cream.

Otago and Southland

At a meeting in 1881 of the Taieri Agricultural Society in Mosgiel, chaired by Donald Reid, one speaker advocating a local dairy factory referred to sending milk regularly by train as far as Dunedin on a contract basis for 6½ pence per gallon for some years, while another, with first hand dairying experience in the United States, considered that such a local factory would not then represent a paying proposition. In 1882 the local newspaper, *Taieri Advocate*, continued and took up the issue, claiming that "... it would be much to the farmer's advantage to be able to

New Zealand Farmers Dairy
Union, manufacturers of Black
Swan butter.

send his milk to a factory on the Taieri instead of having to send it by train
to Dunedin and elsewhere." Plant and machinery for a dairy factory to
handle milk from 500 cows would have cost an estimated £700 or £800,
while local dairy expert, Thomas Brydone, considered such machinery
could be largely manufactured in Dunedin.

In 1888 the newspaper renewed its campaign for a butter and
condensed milk factory, in view of a tariff that virtually excluded any
imports of condensed milk into New Zealand. It also reasserted the
superior quality of factory or creamery made butter, which justified the
higher price commanded. For example the Henley Dairy Factory was
selling its butter for 1/3d per pound, fully 25% more than other Taieri
farmers were receiving for their farmhouse product. The newspaper's
editor asked "Why do not our Taieri farmers keep pace with the times
and go in for dairy factories and co-operation?"

Four months later the first Mosgiel butter factory was opened, not as a
co-operative but as a result of private enterprise, the dairy proprietors
being two well-known local merchants, J and R Cuddie. The manager of
the Henley factory inspected the new dairy premises and described them
as being as good as any he had seen. It was proposed to pay suppliers 3d
per gallon with all skim milk returned to them. On 2 November 1888,
Messrs J and R Cuddie began operations with a satisfactory milk supply

Cuddie brothers early Otago
butter factory.

for their butter factory. The *Advocate* recorded the opening, adding that
as soon as suppliers recognised the advantages of a cash market for their
milk, the proprietors would have all the milk their factory could
process.

By the following August the demand for the Cuddies' butter exceeded
supply and orders had to be repeatedly refused so that a further West
Taieri factory was proposed. Cuddie's was the first dairy factory in the
South Island devoted solely to butter manufacture while in 1891 it
appears to have been the first in New Zealand to install a combined churn
and butter-worker and one of the first to use a Hall refrigerating unit.
Whereas the Edendale product had been shipped in bulk, Mosgiel
became the first factory to forward its butter weighed, wrapped and under
its own brand.

The Cuddie brothers, as storekeepers, were aware of having to deal
with a glut of farmers' butter for which there was little demand and
conceived the idea of buying surplus milk from farmers to convert into
factory butter for export. An early report states: "The farmers did not
take kindly to the innovation at first, but, in the end found it to their own
advantage to support the local butter-factory." The proprietors exported
a portion of their butter to Australia, and sent part of it to London. The
first shipment realised about 10d per pound. After a time several
skimming stations were established to supply cream to the central factory
at Mosgiel. Expert advice was not available, yet the venture was a success.
The staff was small — a manager and one or two labourers.

In 1892 the factory was sold to the Taieri and Peninsula Milk Supply
Company which had begun butter manufacture in addition to its original
milk and cream supply for Dunedin. The Mosgiel factory eventually
became but one of a network of some 58 "T and P" creameries and
skimming stations covering much of the provinces of Otago and

Edendale dairy factory.

Southland, operating until the advent of home separation between 1916 and 1923 when butter operations were centralised at Oamaru and Dunedin.

While the experience of the Mosgiel factory establishment was repeated elsewhere, that of the Edendale factory was a unique instance of corporate activity in the dairy industry's development for it was built as a result of direct investment by the New Zealand and Australia Land Company. As a result of the survey of the varying profitability of the Company's several South Island estates, it was decided to subdivide and sell the Edendale Estate in Southland. To increase the viability of such subdivisions and make them much more attractive to would be purchasers, the Company built a dairy factory, based upon plans of a similar plant visited by William Soltau Davidson at Ingersoll in Canada, and also purchased 300 cows to initiate a milk supply and secured milkers.

The construction of the dairy factory was supervised by Thomas Brydone. The factory operation was actively supervised by the estate manager, RM McCallum, with the work of the factory being carried on by George Inglis, his wife and daughter, and an assistant who saw to the pigs. The second factory manager was Joseph Wood, followed in turn by Thomas Scoullar.

Milkers for the Company's herd were expensive and difficult to secure at first, but later land purchasers and tenants on the Edendale subdivisions became the principal milk suppliers. Dairy cows were milked by women and boys in sheds at 1 penny per cow, but the factory also received milk from farmers with their own farms and herds. There were also 150 Berkshire pigs fed on factory whey.

The factory was reported to have cost £1200 but it also gained the Government bonus of £500 in its second season of operation. The same

Company also initiated the first shipment of frozen meat. The Edendale factory was equipped for both butter- and cheese-making, like most other early dual plant factories, although it initially concentrated on cheese-making. Butter-making, on any large scale, was not attempted for several years. At the pioneer factory butter was being produced before mechanical separators were in common use so the procedures adopted for separating the cream from the whole milk were those of the traditional, old-fashioned, pan-skimming gravity-setting system.

In 1890 Newman Anderson was brought out from Denmark to assist at the factory as an expert butter-maker, together with Danish butter-making plant, including two power-driven Holstein churns each with a butter capacity of 150 lbs and also a power-driven rotary butter-worker.

The New Zealand and Australia Land Company successfully operated the Edendale dairy factory until 1903, when, as a result of the Government's acquisition of the estate for closer settlement, a new co-operative dairy company was formed from the earlier proprietary enterprise. Subsequently the Company built branch factories at Brydone and Menzies Ferry.

The Taieri and Peninsula Milk Supply Company was another corporate proprietary enterprise, operating on a very large scale and ultimately drawing milk supply from an area extending from western Southland to northern Otago. The Company was originally set up by WJ Birch, TO Stokes and J Johnston in order to supply milk and cream to Dunedin and its suburbs. In 1884 the Company began with the issue of 3,000 shares at £1 each, the first premises comprising a small dairy factory and dwelling at a total cost of £600. The Company at first was able to take only a proportion of the milk supply from each of its suppliers who had to dispose of the balance of their milk and butter at most unsatisfactory prices. In 1889 a butter-plant was added and the concern converted into a butter factory, with the first of the skimming station network at Sandymount and Mihiwaka in 1893. By 1900 the business embraced a total of 30 skimming stations, extending from Tokarahi, 80 miles north of Dunedin, to Edendale, 90 miles to the south. The 1884 turnover of £5,000 had grown to more than £110,000 by 1900. At its greatest extent the Company included a network of almost 40 skimming stations and later made a quantity of cheese too. With the growth of co-operation, many of the skimming stations were taken over by suppliers to set up co-operative dairy companies. Some of these converted to cheese production, although with the rapid adoption and spread of the home separation system, many were converted to butter production.

South Auckland and Waikato

One of the earliest dairy proprietors to make significant contributions to the rise of dairying in the Waikato district was Captain J Runciman, farming at Hautapu near Cambridge. He was among the first dairymen to import American manufactured dairy plant and machinery and, as early as 1878, was making cheese at a small factory located on his own farm, using the milk supply from both his own and his neighbours' dairy herds.

But the major thrust from proprietary interests in the district was to be associated later with the three proprietors whose efforts laid the basis of the dairy enterprises that ultimately amalgamated to form Waikato's and New Zealand's dairy giant, the New Zealand Co-operative Dairy Company.

In 1883 the dairy enterprise that was eventually to develop into the

New Zealand Co-operative
Dairy Co dried milk plant,
Waitoa, circa 1924.

New Zealand Dairy Association, the largest of the three dairy companies to later amalgamate, was established when the butter department of the Auckland based New Zealand Frozen Meat and Storage Company was set up, with Wesley Spragg both in charge and with a vested interest in it. The department at the time was a trading concern, buying and re-selling settlers' surplus butter which was blended and packed according to market demands. For the large local city trade it was packed in 1½ lb pats while the balance would be put into 112 lb kegs for disposal in the sawmillers' camps or exported to New South Wales. At one time buying operations extended as far as Waitara in Taranaki. The prices paid for butter ranged widely, from as much as 6d or 7d per pound to as little as 2d delivered to the depot! Under this system Spragg's department bought, processed and sold about 300 tons annually, a significant quantity at that time.

A few years previously several efforts to establish small co-operative cheese factories in the districts where butter was collected had failed and factories at Pukekohe, Hamilton, Paterangi and Te Awamutu were subsequently closed. After such an experience few were keen to carry on with similar co-operatively based dairy enterprises. Wesley Spragg, however, recognised the opportunity to improve the quality of dairy processing by taking the whole milk supply from local farmers and handling it along the lines of the factory system already proving so successful in other dairying districts. After discussions with potential factory suppliers, arrangements were planned and completed to purchase milk and manufacture butter, the former cheese factory at Pukekohe being acquired and fitted out with milk receiving vats and mechanical cream separators. A Danish butter-maker was engaged to take charge of the creamery operations at a site on the premises of the freezing company, on reclaimed land near the railway station, for the separated cream would be sent north by rail to Auckland for butter manufacture. Settlers' whole milk for initial separating at Pukekohe was bought and paid for by weight, 2½d per gallon of 10 pounds. Shortly after the operations began, the creamery was transferred south to the Pukekohe factory which, with innumerable additions and alterations, remained in constant use until destroyed by fire in 1923.

Spragg's business acumen was evident when he was able to secure the financial backing of London based dairy produce agents, Lovell and Christmas, who not only provided necessary investment capital for expansion but also served as a London marketing representative when later the New Zealand Dairy Association, having been established by Spragg, absorbed by purchase, Henry Reynolds and Company.

The New Zealand Dairy Association was formed into a co-operative concern in 1901 when the 847 shareholders paid £40,000 for the business owned by the joint partners, Wesley Spragg and Lovell and Christmas. The NZDA at that time had an annual butter output of 1200 tons, of which some 400 tons were consumed locally and 800 tons exported to London. Wesley Spragg remained with the new enterprise as managing director, while the plant itself comprised two central butter factories at Pukekohe and Ngaruawahia, and a network of 40 skimming stations. The new co-operative undertook to build a new skimming station wherever a district could guarantee the daily milk supply of at least 250 cows, while payment to suppliers was based upon the size of the skimming station or the quantity of milk handled.

In 1886 Henry Reynolds built a butter factory on his farm at Pukekura, the first factory in the Waikato built solely for butter manufacture. It was managed by David Gemmell, an American, trained and experienced in butter-making, who was then farming in the Waikato. The original factory had the capacity to process the milk from 100 cows on Reynolds' farm but later this was augmented by neighbours' herds.

Reynolds was responsible for the choice of the original brand name, "Anchor", allegedly because he had such a tattoo on his arm. Butter sent to the Melbourne Exhibition in 1888 under this brand was awarded first prize. Success on both the local and Australian markets encouraged expanded production as a result of increased demand, and additional capital from neighbours was invested in extending the skimming station network to enlarge the milk and cream supply. Subsequently the expanded enterprise was reformed as Reynolds and Company in 1888 or 1889.

As the business grew factories were set up at Ngaruawahia and Newstead with skimming stations at Te Kowhai, Whatawhata, Paterangi, Te Awamutu, Kihikihi, Pukerimu, Hamilton, Waihou and Te Aroha West. In 1891 the company briefly spread its operations into northern Taranaki with a butter factory at Inglewood and skimming stations at Kaimata, Egmont Village and Tariki, but the venture closed down when the Taranaki interests were sold in 1895. In 1896, Henry Reynolds finally sold his dairy interests to Wesley Spragg, to be merged with the New Zealand Dairy Association. The takeover provided for the adoption of Reynolds' own Anchor brand by the enlarged company.

William Goodfellow a hardware merchant, imported for one of his clients, American equipment for the manufacture of butter from separated cream. The client failed to take delivery, and rather than leave the machinery unused he engaged a factory manager familiar with the home separation system of butter manufacture, and a small factory was set up at the back of the Hamilton Horse Bazaar in 1909. At the close of its first season the company was reorganised as a co-operative. It was indicative of prevailing attitudes that suppliers were initially slow to take up the available shares. Only with the inducement of an extra bonus paid to shareholders did the company begin to develop as an active co-operative.

The certificate of incorporation as a co-operative company was registered in 1912 with Goodfellow as managing director. The company grew so rapidly that the small Hamilton building soon became inadequate and a new factory was built at Frankton in 1913. The Waikato Co-operative Cheese Company was formed 2 years later to operate independently but in close association with the Waikato Co-operative Company. To utilise the cream supply available in the lower Waikato and to expand butter output, a new factory was built at Tuakau in 1917 to replace a butter factory operating at Mercer on the Waikato River. Amalgamation with the NZDA was effected in 1919.

Taranaki

Taranaki's early dairy processing industry was dominated by relatively small-scale proprietary dairy enterprises which suggests the source of capital was on an individual, and consequently, limited basis. Many of the early factory proprietors were themselves milk suppliers with some experience in butter- or cheese-making. Factory dairying was the means of attaining a larger volume of production with the associated assurance of a reliable and uniformly high quality product.

Alfred Brake's private cheese factory, set up at Lepperton in either 1882 or 1883, 10 miles north-east of New Plymouth, marked Taranaki's initial entry into factory dairying. Brake himself collected milk from

Crown Dairy Co, Okato, 1901.

nearby farmers for processing, together with that from his own herd, driving an old-fashioned two-horse team on set days each week.

Brake was also active in early dairying in other areas. In 1885, when the suppliers of the co-operative Moa Dairy Factory Company voted to extend into cheese-making as well as butter, he was employed as an early Manager. He was paid £2.2.6d. weekly when only butter was made but £3.6.6d. during that part of the season when the factory, as a dual plant, made cheese as well. Brake also managed the Opunake Dairy Factory Company in its second season in 1886, and assisted SA Breach to install the De Laval separator as well as supervising the equipping of the creamery on Ihaia Road when Breach set up his own enterprise.

Samuel Breach, co-founded the Opunake Dairy Factory Company in 1885 in association with schoolteacher/dairy farmer JJ Elwin of New Plymouth. The early cheese factory was one of the most successful with Thomas Cranswick, formerly first assistant for Captain Runciman's Waikato cheese factory as manager for the first season, and for the second, Alfred Brake, formerly of Moa and Lepperton. Such was the milk supply in the spring flush of 1886 that the factory organised a night shift to ensure it was all properly utilised.

Breach was the largest, single factory supplier and in 1886, withdrew to form his own company. The creamery built on his farm on Ihaia Road was later referred to as the Opunake Creamery. At this time Breach was milking about 150 cows and employing 5 milkers. The power for his De Laval separation was supplied by a single horse. A cool-room was built with a double roof and solid walls, with water for cooling obtained from a nearby stream by means of a force pump.

Breach sent several successful consignments of kegged butter to Australia and England. His first shipment comprised 16 kegs of butter sent to Sydney by boat via Wellington in July 1886, the realised price of 1/4d to 1/6d per pound indicating it was of good quality.

Another proprietor contributing to the early introduction of factory dairying in Taranaki was Thomas Bayly. Apart from Lepperton, all the earliest dairy factories in Taranaki were located south of New Plymouth, but in 1887 dairy development to the north was boosted significantly when Bayly set up a butter factory in Waitara Road. Bayly's own milk supply was augmented by milk purchased from neighbouring farmers at 2¼d per 10½ lb gallon during the first season when the factory was supplied by as many as 300 cows. Two Burmeister and Wain separators, with a combined capacity to handle 160 gallons of whole milk per hour, were utilised. The butter was exported in wooden kegs in 60 lb and 100 lb lots.

In equipping Bayly's factory for butter-making, a locally built butter churn was installed, thought to have been one of the largest in New Zealand at that time. It was so large, in fact, that a hole had to be cut in the existing wall to install the churn. The wooden churn was steam-operated with barrel and stand that stood over 5 feet high, while the capacity of the barrel itself was just over 400 gallons, handling approximately 350 lbs of butter in a single churning. Although large by the standards of the times, such a churn is miniscule compared to those used today.

Taranaki's most celebrated dairy proprietor was Chew Chong who arrived in 1870 to work as an itinerant pedlar of children's toys, trinkets and household needs. His two main contributions to dairying were establishing the export fungus trade and factory dairying.

Chew Chong travelled regularly through some of Taranaki's recently cleared and grassed bushlands and had noticed the edible fungus Jew's Ear growing on the stumps and logs of tawa, pukatea and particularly mahoe,

Jew's Ear, an edible fungus which provided much needed cash income for many Taranaki farmers.

that littered the formerly forested countryside. The fungus was similar to one highly prized by the Chinese. The cash received for the fungus meant much to struggling settlers and for many families may have been the difference between selling up and holding on to farms for better times. Surplus butter had to be bartered with the storekeeper for supplies but fungus once gathered meant immediate cash in hand. After arduous labour, butter often realised little more than 4d a pound, and that had to be taken in provisions. Fungus, growing freely in the countryside and easily gathered by the younger members of the family, fetched 3d a pound. Contemporary reports suggest that many settlers frequently found it difficult to find the money needed to pay annual local body rates and similar charges, so although at first sight there would appear to be little connection between fungus and dairy produce, in Taranaki in the 1880s the relationship was very close indeed.

In 1874 Chew Chong had shipped consignments of milled farmhouse butter to Australia but as an export venture such butter had proved notably inconsistent all too often losing money. Later in 1884 Chew Chong left his store in New Plymouth to set up in Eltham. He again bought farmhouse butter for milling and resale but being unable to dispose of it in New Zealand tried a shipment to the English market with dire consequences: Chew Chong paid the farmers 4d per pound for their farmhouse butter for milling, paid 3d per pound steamer freight, only receive 70 to 75 shillings per cwt, or 7½ - 8d per pound, without considering further incidental costs in handling the shipment, and London market reports were far from encouraging.

Rather than depend on unreliable supplies of farm butter, Chew Chong built his own butter factory in Eltham in 1887. It was called the Jubilee factory in honour of Queen Victoria's celebration. He later erected a network of skimming stations at Hunter Road, Te Roti, Mangatoki and Rawhitiroa with Burmeister and Wain separators in each.

In 1889, at the Dunedin Exhibition, Chew Chong's creamery butter entry was awarded the silver cup for the best half ton of butter packed ready for export. He was not himself a practical butter-maker and the winning entry was the work of his factory manager, Sydney Morris,

New Plymouth.

November 1910.

Mr Chew Chong,

Dear Sir,

We the undersigned Settlers in the Taranaki District wish to place on record our appreciation of the services which you have rendered to your adopted country. Your having on your arrival from China in the early sixties entered into the export trade in Fungus was the means of saving many a family from want and penury in the early days of settlement in the district by causing a circulation in our midst of over half a million of Foreign money and this the outcome of legitimate trade and not from loan. When later you entered into the Butter business being almost a pioneer in Factory Manufacture you led the way into what has become the mainstay of the district and helped to develop an export which materially assists in the prosperity of the Dominion. That this industry passed from private into co-operative hands and thereby caused you considerable pecuniary loss is on your account to be deplored but cannot be traced to a want of business acumen nor to a want of desire to assist your fellow settlers on your part. We wish that we may for many years constantly meet you in our midst and that you may enjoy the rest to which your years and exertions entitle you.

formerly of Moa. Later, when many of his suppliers formed their own dairy co-operatives, Chew Chong resented the fact that he was virtually driven out of the industry as his milk supply rapidly dwindled and his factory's continuing existence was threatened with the spread of farmer owned and operated co-operatives.

Among his other contributions he invented a butter-worker and was one of the first to make use of an "impress" butter brand. In 1889 he installed a Hall's refrigerating machine in his Eltham factory, the first freezing-machine installed in any New Zealand butter factory. He was also a shareholder in the Egmont Box Company in which 14 of the 17 shareholders were dairy factories. He adopted a form of labour on some of his farms very similar to sharemilking, although the actual introduction of sharemilking as such was attributed to the Henley Land Company on the Taieri in 1884.

One of the few larger corporate proprietary dairy concerns in Taranaki was the Crown Dairy Company of Newton King and JC George, later joined by R Cock and O Samuel. Like so many proprietaries, the Crown began its dairy operations by taking over and revitalising several local failed dairy co-operatives — Otakeho, Manaia and Opunake. Crown not only took over the operation of the closed dairy factories but also provided far-sighted financial assistance to the struggling settlers to purchase various dairying prerequisites to improve their farms and herds,

Opposite page: A well earned testimonial to Chew Chong.

Below: Henry Reynolds' butter factory.

the loans to be repaid by regular deductions from milk cheques. Such assistance assured the Company of expanding milk supplies which in turn meant reduced per unit manufacturing costs and ultimately, better returns for local suppliers.

By 1897 the Crown Dairy Company had developed its own distinctive pattern, operating 18 branch factories and 2 skimming stations in Taranaki, and a factory in Woodville. Apart from a brief incursion by an Auckland dairy enterprise, the Crown Dairy Company continued to dominate the development of factory dairying in northern Taranaki from 1890 to 1900. It closed in 1903 selling out as a result of the rise of dairy co-operatives.

Manawatu

While the 1880s were largely dominated by proprietary dairy enterprises, the 1890s witnessed the growing importance of dairy co-operatives. On the other hand, while in other dairying districts early dairy co-operatives had failed only to be taken over, re-opened and operated successfully as proprietary dairy factories, in the Manawatu, dairy proprietaries and co-operatives continued to operate side by side, competing with one another for the available milk supply, with one notably bitter rivalry between two of the larger concerns.

The first dairy factory in the Manawatu was a private cheese-making venture begun by James Skerman near Newbury about 4 miles from Palmerston North. Skerman and his family arrived in the Manawatu in 1877 and developed a bush-covered, 350-acre section on what is now known as Skerman's Line. In the first year 20 acres were cleared and dairying began with a herd of about 10 cows, making butter to sell to stores in Feilding and Palmerston North. In about 1879 or 1880 the Silverleys dairy factory was built on the farm, the first in the district, making cheese for sale in the same markets. The dairy operation was fairly small at first, handling little more than 3,500 gallons of milk in the first season but by the seventh season the supply had increased to as much as 45,500 gallons.

During the 1880s, the Silverleys cheese factory was moved to the larger building at the junction with Rangitikei Line, being sold in 1907 to the New Zealand Farmers' Dairy Union, who continued cheese-making. In 1911 the factory was purchased by the Newbury Co-operative Dairy Company which continued manufacturing cheese under the old brand label until the 1940s when it became a milk collection depot for the Glaxo milk powder factory at Bunnythorpe.

The second dairy factory venture, at Karere, had its origins with Johannes Monrad, noted for the dairy appliance catalogue issued in 1883, advertising hand-separators. Monrad had imported from Denmark the first centrifugal mechanical cream separator in the Manawatu in 1883, one of the earliest in New Zealand, to demonstrate to local dairymen and encourage them to set up their own factory to utilise the latest invention. A butter factory utilising the milk from 300 cows was set up by suppliers at Karere in 1883-84. It produced mainly butter, selling milk, cream and butter in Palmerston North. Like so many other co-operatives Karere was not a success and soon went into liquidation. It was sold in 1886 to a private partnership of Buick & Co and converted to cheese-making and bacon curing, but finally closed in 1890.

William Wescombe Corpe arrived in the Manawatu in the early 1880s and set up a small store and steam saw-mill in Makino. Like so many other storekeepers he bought small amounts of surplus butter from local farmers in the Feilding district to blend before selling locally or elsewhere

in the Wellington district. To supply the butter factory built at Makino in 1888, Corpe bought bulk milk supplies from a wider area. His butter was marketed under the brand names "Safe", "Tip-Top", "K", "Kamoro" and "Milkmaids".

WW Corpe's Makino butter factory.

Some butter was exported by ship to England, after being railed to Wellington, and was normally placed on board in the cool-room with the cheese. On one occasion in the early 1890s, when there was no space in the cool-room for the butter kegs, rather than leave the butter to spoil, Corpe agreed to put it in with the export meat in the freezer compartment. Arriving in first class condition, the butter commanded a premium price and was the first of many shipments of frozen butter.

Corpe's suppliers included Campbelltown farmers and in 1893 he opened a separate factory there to process milk locally. The separators and manufacturing plant were steam-powered. To cope with all the available milk supply, the separators were operated from 6.00 am to noon, farmers delivering milk in horse-drawn vehicles, with many impromptu chariot-races on the way to the factory and wagons often left stuck in the mud or the ditch! There were 22 suppliers, mostly with herds of about 20 cows, and the largest individual supply was about 50 gallons daily.

Soon after the opening of the new butter factory, butter prices slumped badly. Corpe, £1,500 in debt and facing bankruptcy, decided to sell up in 1894, offering the suppliers the alternative of factory closure or purchase for £1,100 to form the Campbelltown Co-operative Dairy Company.

The new co-operative, later named Rongotea, began in 1895 with 10

Motorised milk/cream can collection for the Taieri & Peninsula Milk Supply Co.

suppliers and 37 lbs of butter made in the first week. Butter was sold in bulk as "Milkmaid" with pat butter as "Makino". In the first year the number of suppliers increased to 60, and over £7,000 worth of butter was sold. Initially suppliers had received 7d - 9d per pound of butter.

In 1893 suppliers formed the largest co-operative for the Wellington-Wairarapa-Manawatu district, New Zealand Farmers' Dairy Union, based in Palmerston North. In 1894 the cream supply for Palmerston North was received from skimming stations in Whakarongo, Ashhurst, Pohangina, Rangitikei Line, Foxton and Sanson, but in the same year its manufacturing centre was transferred to Thorndon in Wellington with a network of 13 skimming stations for cream supply as far north as Tokomaru. Distance proved a problem, however, and in 1899 it returned to Palmerston North, expanding its operations to a peak of 2 creameries and 36 skimming stations. The first manager was WJ Birch, founder of Taieri and Peninsula Milk Supply in Dunedin and later superintendent of the Crown Dairy Company in Taranaki.

Joseph Nathan and Company had financially assisted various dairy enterprises both proprietary and co-operative, in the Manawatu district. This had included the New Zealand Dairy Farmers' Union, which not only received assistance by way of loan money but also proprietary butter export marketing facilities. This stable dairying relationship between Joseph Nathan and the NZFDU was upset suddenly in 1900. As a result of misunderstandings arising from an NZFDU decision to repay a substantial loan, Nathans countered by setting up in opposition to the co-operative, beginning butter manufacture and marketing under the company's brand "Defiance", as well as "Black Rooster" (NZFDU's brand was "Black Swan"). Nathan's Defiance Dairy Company expanded rapidly to include 2 creameries and a network of 40 skimming stations, mostly located adjacent to or close by the existing NZFDU skimming

Edendale, factory interior circa 1908.

stations. Nathan's efficient butter-making and higher milk and cream payout attracted suppliers from the co-operative, the latter being notoriously inefficient in its operations.

But as a proprietary concern Nathan's role went much further. In 1901 milk powder production was begun in the Makino factory, formerly acquired from WW Corpe (and later the Cheltenham Co-operative Dairy Company), and a second factory was set up in Bunnythorpe in 1905, later to be developed under the "Glaxo" brand. Nathans demonstrated the business acumen and technological expertise to survive and later pioneered yet another dairy product, casein, in 1911 at Aramoho in Wanganui.

Wairarapa

While other dairying districts illustrate the initiatives of dairy proprietors in promoting factory dairying the Wairarapa provides an excellent insight into the early reactions of small farmers. The small dairy farmer was preoccupied by the problem of subsistence. His behaviour was governed by large measures of independence and self-interest which left him very sceptical of the benefits to be gained by co-operative dairying. This stance, characteristic of many small dairy farmers and their communities, contributed greatly to the two underlying problems inherent in attempting to establish factory dairying — adequate milk supply and capital investment.

Given the widespread shortage of cash in hand the disinclination to make any financial commitment to a dairy factory is understandable. The reluctance to provide a regular supply of milk, however, had its genesis, to some extent, in cupidity. The supply of milk to factories for cheese-making necessarily limited the amount available for farmhouse butter manufacture which in turn led to an inability to satisfy market demand.

The resultant increase in butter prices made farmers unwilling to supply factories with milk and many would do so for only part of the season, in order to be able to take advantage of the shortages created and the inevitably inflated prices. These attitudes meant that the establishment of dairy factories, particularly in areas of marginal support such as the Wairarapa, was a prolonged exercise and success in the enterprise was more often than not in spite of the farmers rather than because of their support.

The first dairy factory in the Wairarapa was a proprietary venture instigated by Messrs Gilpin and Pardon, owners of a small retail business in Featherston. In 1880 they established a small cheese factory which processed a daily milk supply of only 250 gallons. Cheese was the logical product to manufacture since there was little made at the time and there was thus a steady demand, not only locally but also from the nearby urban market of Wellington.

Gilpin and Pardon's factory produced cheese for two seasons and enjoyed marked success. In their third season the proprietors announced their intentions of accepting all available milk offered to them by local suppliers and of moving to larger factory premises. Why such plans were not carried out and why, after 2 years of successful operation the factory suddenly closed, is not fully known. The most likely reason is that the local milk supply was withdrawn in order that suppliers could take advantage of the high prices being offered for farmhouse butter. Farmers' attitudes may have played a pivotal role. Many grossly overestimated the value of their milk arguing that it was preferable to feed it to calves and pigs rather than accept 3½d a gallon for factory supply.

10

CREAMERIES

In relation to butter-making, the first significant variation in the factory system of dairy processing was the change from one locally supplied dual plant factory to the creamery system of butter manufacture. The creamery system comprised a single, centrally located creamery or butter factory, and a network of skimming stations which supplied cream. Whole milk was delivered daily by local suppliers to the skimming stations where it was mechanically separated and the cream forwarded regularly to the creamery.

Origins and Organisation of the Creamery System

The first dairy factories had usually drawn a regular and adequate milk supply from local suppliers farming in an area not too far distant from the factory itself. Such an assured milk supply was a precondition of setting up a new factory. At Edendale, for instance (the first dairy factory as such), in the absence of existing smallholdings and dairy herds, the New Zealand and Australia Land Company set up its own herd to assure an initial milk supply and also attract other small suppliers. As the factory system spread difficulties arose in ensuring an adequate and regular milk supply.

The creamery system was essentially a spatial or areal response to the problem of factory establishment and operation, or more specifically, to several particular and related aspects of factory milk supply. These were the limited distances over which whole milk could be effectively and satisfactorily carried by the prevailing form of transport of the day. Other factors were the generally small size of early dairy herds and the overall relatively low livestock numbers. Given these constraints the problem was one of ensuring an adequate milk supply. Thus if the existing 4-5 mile milk supply area had too few cows the solution lay in extending the area itself. Distance precluded the continued expansion of an existing supply area so the alternative was to set up several additional but related small supply areas within which the dominant forms of transport could function readily. The minimum supply for each creamery was estimated to be 300-400 cows.

The ensuing competiton for milk explained much of the duplication of skimming stations in some dairying communities by rival dairy factories anxious to consolidate or expand supply. In most developing bush dairying districts it was rare to find an adequate number of dairy cows close by the creamery and so skimming stations were used to expand the

The Mercer creamery which relied on launches for milk collection.

area from which milk supply was drawn. A further factor was that, given the form of early mechanical separation with large capacity and plant that was heavy to operate, a considerable bulk or volume of whole milk had initially to be moved from the farm for separating. By the early decades of the 20th century, when the creamery system began to be superseded by the home separation system, farmhouse or hand separators had become increasingly common. Similarly, motorised transport replaced traditional horse-drawn vehicles significantly extending the cream collection area.

As butter manufacturing capacity grew (and cream demand increased proportionately) an existing network of skimming stations could be extended to expand the supply area, and similarly, stations could be closed or converted to cheese-making as demand eased or declined. Extending or contracting the network of skimming stations was a convenient way of increasing or reducing numbers of suppliers, while still operating and functioning within the limitations imposed by the available transport.

In addition to problems of adequate supply and distance of delivery, there were also processing considerations. The 1880s saw increased availability and adoption of mechanical separators such as the De Laval and Burmeister models. This new form of milk separation rapidly displaced older and more traditional methods. But the new mechanical separators introduced constraints of their own. Their size, increased capacity, the heavy nature of their mechanical operation, and their widespread utilisation of steam or water for motive power, meant they had, of necessity, to be installed in more permanent buildings than had hitherto been used. At the same time, they had to be as close as possible to the source of milk supply.

Whereas the creamery was large in size, as it also included plant for butter manufacture, the skimming station was a unit on a much smaller scale, fitted up only with scales, measuring and storage tanks, mechanical separators and motive power by steam engine or water turbine. It was so

designed that the total separation process functioned on a gravity flow basis from stage to stage. The freshly delivered can of whole milk was hoisted onto a loading platform for emptying into the tank and weighing before the milk flowed down through the separators. The skim milk was finally returned to the supplier's milk can at a lower point in the system.

Top of page: Dairy factory, Stratford, circa 1902.

Above: Eketahuna dairy factory built by the New Zealand Dairy Union 1903.

Above: Delivering whole milk to the factory and receiving the skim.

Right: Milk delivery and mechanical separation at the skimming station circa 1908.

Throughout the late 1880s and the early 1890s, the New Zealand dairy industry expanded and as it did so, the creamery system became more widely adopted. By the 1893-94 season, when by government legislation all dairy factories had to be registered, incorporating registration numbers in the brand designs used, there was an overall total of 62 creameries. The system reached a peak around 1904-1908. In its 1904 *Annual List of Creameries and Dairy Factories*, the Dairy Division of the Department of Agriculture listed 217 registered creameries supplied by 446 skimming stations, and a total of 983 registered dairy premises.

Functioning Units and Features of the Creamery System

The creamery system operated for approximately 40 years in all. During that time its regional distribution was somewhat uneven. The one common feature, however, was its characteristic spatial or areal patterning as a response to the distance and supply relationship, and the dependence upon horse-drawn vehicles for delivery to the factory.

After early morning milking the large milk cans were loaded onto all kinds and varieties of primitive, mainly horse-drawn vehicles. The whole milk was delivered to the skimming station as early as practicable. The daily delivery of morning milk was a feature of the creamery system, suppliers driving from near and far, queueing up at the factory with their vehicles to deliver and then collect a share of the skim to take back to the farm for feedings pigs or calves. Much time, which might have been utilised more usefully was spent by the suppliers patiently waiting in line for their milk cans to be hoisted up and the contents emptied and separated. Consequently, the more distant suppliers often raced their horses and vehicles to try to secure a front position, to the detriment of the quality of the supply.

The vehicle line up and the attendant time for suppliers waiting to complete delivery did, however, provide daily opportunities for them to exchange ideas and communicate usefully with one another. Such regular meetings played a significant role in the spread of new ideas among a community.

Once the milk had been separated and the skim returned, the cream was forwarded daily to the creamery by whatever form of transport was available. Although this was normally by road or rail, in some instances it was by water transport along rivers or around harbours and coasts.

Network Patterns and Factory Locations

The spatial patterning and size of each skimming station network varied within and between districts. Some areas were dominated by relatively small networks. In Taranaki, for example, creameries were characteristically supplied by networks of less than five skimming stations and a large proportion, it might be noted, with only one or two. Other areas were dominated by only a few butter-making creameries, each of them having an extremely large and widespread network of skimming stations to receive whole milk deliveries.

Several explanations have been offered to account for the areal or spatial contrasts of the system between one district and another, particularly where different networks dominated. It is of special interest to note that in both the Taranaki and South Auckland-Waikato areas the earlier overwhelming dominance of small- and large-scale networks respectively has persisted in contemporary patterns with numbers of characteristically small or medium sized enterprises in Taranaki, and South Auckland-Waikato dominated by the country's largest single giant dairy enterprise, the New Zealand Co-operative Dairy Company. Some have suggested that dairy patterns are influenced by and reflect their contrasting physical environment. Thus river valley situations like the Waikato tended to integrate, link and produce more cohesive networks, while a mountain mass like Egmont tended to separate and divide one from another, creating very strong parochial sentiments and associations that mitigated against co-operation on any broad regional basis. Perhaps more significant were the sources of finance. Larger dairy enterprises required more capital investment and as such often derived capital from a single external source that facilitated the organisation of large-scale, centrally operated and controlled dairy enterprises over a relatively wide area. On the other hand, supplier-dominated co-operatives depended heavily upon local investment and so tended to remain small, retaining most of the significant decision-taking and policy-making authority within the control of the local dairy community.

Apart from the contrasting patterns and associations of creamery networks, there were also contrasts in the specific locations and sites of dairy units. By the turn of the century a dairy cow distribution pattern was beginning to emerge nationwide. Most dairy cows, and consequently milk supply sources and dairy factories, tended to be concentrated on the more level and rolling land below 1,000 feet above sea level, and even more especially on lowlands of less than 500 feet above sea level.

Within the bush and fern dairy lands specifically preferred locations and areas emerged as factories and farms began to concentrate within a district. During the period when the creamery system was dominant, specific factory locations reflected the historical circumstances of transportation developments. In Canterbury, Otago and Southland,

Skinner Road, Stratford.

railway networks were developed at an early stage pre-dating the spread of the early dairy factory system. Consequently, in these areas both skimming stations and creameries were located so as to facilitate the loading and transport of cream by rail.

Railways developed later and more slowly in the North Island. The main trunk railway line, south from Auckland and north from Wellington, was not joined until shortly after 1900. As a result, railways in the North Island were less important factors in the location of dairy factories.

Road networks and intersections played a major role in Taranaki factory locations in the formative years of the dairy industry as did bridge-points along the "circular" roads necessitated by the numerous mountainslope streams. Many of Taranaki's early, narrow surfaced roads have survived. A feature of early dairy factory location in Taranaki that also survived until recent decades was the siting of small dairy factories, both butter and cheese, every 5 or 6 miles along the main roads.

Further north in both the Waikato, Thames-Hauraki and Kaipara

York Road creamery, Midhurst Co-operative Dairy Co.

districts, water transport was all-important and riverside and harbour or coastal locations were chosen for dairy factories. Rivers like the Waikato and Waipa, and the Waihou and Piako, were significant location factors.

Decline and Change in the Creamery System

From its peak period of 1904-08 the creamery system slowly declined and was gradually superseded by the farmhouse separation system. By the late 1920s and early 1930s, the latter was the dominant dairy production system.

The gradual changeover of individual dairy factories and companies from the creamery to the home separation system can be determined by examining the receipts of raw material, whole milk or cream, for processing as listed on the annual balance sheets and the changing proportions of each received. With direct delivery to skimming stations for separation, whole milk receipts made up the largest proportion but with the growth of the creamery system, cream receipts began to represent a larger proportion until eventually only cream was received.

Thus, cream and skim milk were initially separated by traditional farmhouse gravity setting methods, then by off-farm factory mechanical separation in the creamery and finally by on-farm farmhouse or home separation in the milkshed. With the advent of tanker collection in the 1950s and 1960s cream separation once again reverted to the large central factory.

11

GOVERNMENT

AND THE FARMER

T he first real evidence of the Government's practical interest in the dairy industry was an offer of a bonus, as advertised in the *New Zealand Gazette* of 18 May 1881, under the heading of "Bonuses on Colonial Industries". The bonus offered was £500 for the first 25 tons of butter or 50 tons of cheese to be exported. There was a stipulation that the produce had to have originated in a factory operating on "the American principle". This system was akin to a form of producer co-operative and had many features significantly different from those of the butter and cheese dairies of the United Kingdom.

The bonus in question was awarded to the Edendale dairy factory for its second season's output in 1882-83, when it successfully exported the stipulated 50 tons of cheese. It is of interest to note that the small experimental dairy consignment from Edendale in the *Dunedin* in February 1882 had been butter not cheese.

In May 1881 when the bonus was advertised there were only a few dairy factories in New Zealand. The earliest recorded factory was at Highcliff and had been established a decade earlier. However, the Government of the day was evidently favourably disposed towards the adoption and spread of the factory system of dairy processing and manufacture.

Although the factory system was widely advocated, both in New Zealand and overseas, refrigerated transport had not yet been developed. Efforts in this direction were, however, being investigated overseas and it was felt that the offer of a substantial bonus might provide the impetus needed for shipping developments that would facilitate the development of a dairy export trade.

Government Dairy Lecturers and Dairy Instructors
Following the offer of a bonus and the opening of three new dairy factories in 1882, together with the prospect of additional factories in 1883, there was a need for assistance and advice on factory design and equipment. The Government was approached, and its initial move was to appoint William Bowron to prepare a report on the general dairying situation throughout New Zealand. As a result, in 1883 the pamphlet *Observations on the Manufacture of Cheese, Butter and Bacon in New Zealand* was published. Bowron was most enthusiastic about New Zealand's suitability for dairying, and the prospects of an unlimited market in Britain for dairy produce. The report also emphasised the advantages of

factory manufacture compared with traditional farmhouse methods.

Bowron's appointment in 1883 marked the beginning of a Government supervision and advisory service. Five years later an instructional service was established. RM McCallum, former manager of the Edendale Estate, was appointed by the Minister of Lands to travel the Colony lecturing on the operating and establishment of dairy factories. His opinions and findings were presented in a booklet, *Report on Dairy Factories in New Zealand*, together with other papers on the subject.

In 1889 a permanent and practical instructional service was initiated with the Government appointment of John Sawers as the first Chief Dairy Expert. Sawers was one of four brothers from Scotland, all of whom were experienced and skilled cheese-makers. Born in 1866 at Mains-of-Park Farm, Glenluce, Ligtownshire, John had developed an early interest in livestock, dairy farming and the manufacture of dairy produce. In 1881 he had helped his father manage White Hills and Broughton Mains dairies in Ligtownshire, each dairy having about 100 cows. He managed the latter dairy until 1884. Sawers arrived in New Zealand in 1885 and was appointed manager of the Waiareka Dairy Factory at Oamaru. In 1888 he became lessee of that factory and also of the Flemington Dairy Factory. In August 1889 he was appointed Chief Dairy Expert.

In the early 1880s, JB Harris, one of the foremost cheese exporters in Canada, was engaged by the South of Scotland Dairy Association to introduce the most recent Canadian cheese-making methods to Scotland. Harris had originally come from the United States where he had been trained by Professor Arnold, a leading American expert, who was a highly experienced cheddar maker, instructor and organiser. Exposure to Harris meant that Sawers was familiar with the latest processes and most up-to-date practices in the making of cheddar cheese. He also espoused the co-operative system of dairy factory operation and organisation. In 1890 the first of the dairy associations, the New Zealand Middle Island Dairy Association, was formed as a result of the efforts of John Sawers. He had devoted much time and effort to promoting the advantages of some form of co-ordinating organisation to link the various branches of the industry and, particularly, to facilitate joint efforts in matters such as shipping and marketing. Thomas Brydone of Edendale and the New Zealand and Australia Land Company also played a conspicuous part in founding the association and was made the first chairman.

Apart from taking care of the general interests of the dairy industry, the association included in its more specific objectives the improved manufacture of dairy produce and the facilitation of the best and most effective means of transit and disposal of such dairy produce. The Government looked favourably on the association from the outset and after its first season of operation provided an annual grant or subsidy of £200 towards working expenses. In June 1894 with continued Government support, Sawers persuaded those dairymen who attended a meeting in Hawera, to form a similar association, the North Island Dairymen's Association of New Zealand.

The northern body finally adopted the name the National Dairy Association of New Zealand (North Island) and the original southern body became the National Dairy Association of New Zealand (South Island). The latter, however, in 1909 changed its name yet again to the South Island Dairy Association of New Zealand Ltd, and the two organisations were referred to as NZDA and SIDA respectively.

Late in 1891 the Government's continued commitment to promoting instructional and educational services in the dairy industry was clearly demonstrated by the appointment of a further dairy instructor to assist

Opposite page: Recommended plans for skimming stations and creameries.

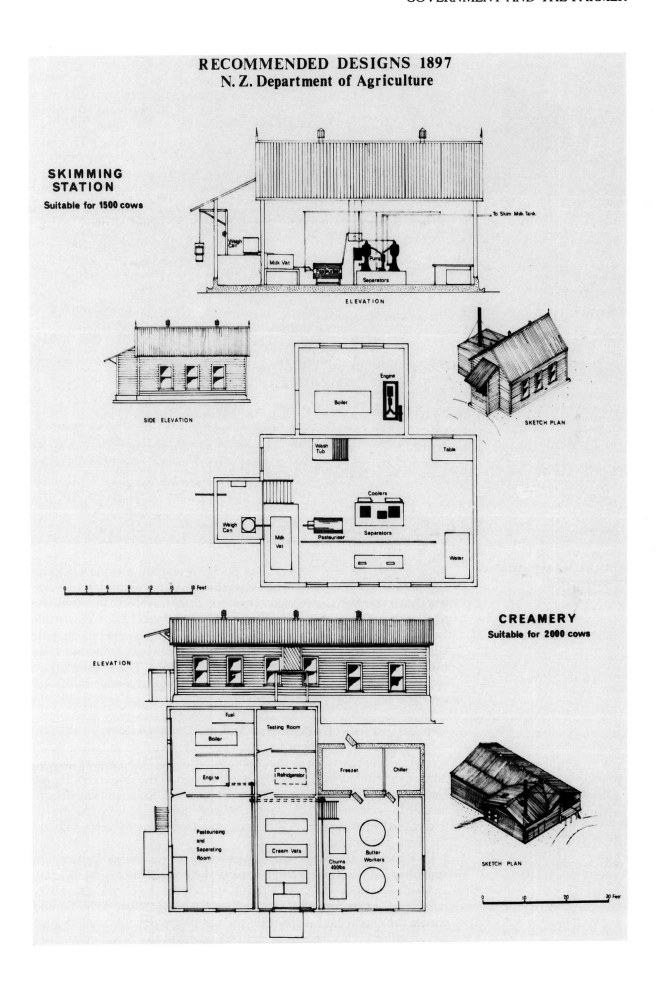

RECOMMENDED DESIGNS 1897
N. Z. Department of Agriculture

SKIMMING STATION

Suitable for 1500 cows

CREAMERY

Suitable for 2000 cows

Mr Linn's herd of hand picked
Jersey cows 1903.
Photo St Clair.

John Sawers. While Sawers was a cheese expert, the new Danish appointee, Carl W Sorensen, had specialised in butter manufacture. After less than a year Sorensen resigned to set up dairy factories. An interesting observation at that time was his being refused entry to 12 factories, while in another instance it was recorded that he was not wanted to examine either the butter or the cheese! At this time the newly built factories had extreme difficulty in filling the manager's positions. There was a severe shortage of applicants who had the appropriate training, skills and experience.

The Department of Agriculture and the Dairy Division

In 1892 the Government formed a single department to look after all agricultural matters. Before this, administration had been effected by the Stock Department and the Agricultural Branch of the Lands and Survey Department. This agricultural section of the Lands Department had been formed in 1886. In 1892 the section's Dairy Instructors were transferred to the new Department of Agriculture which took over the responsibility for dairy instruction.

JD Ritchie, a Canadian, was appointed first head of the Department and immediate stops were taken to place the infant dairy export industry upon a sounder footing. As a result, in October 1892, the first Dairy Industry Act was passed "... to regulate the manufacture of butter and cheese for export and to provide for the purity of milk used in such manufacture". This legislation dealt mainly with the branding of dairy produce intended for export. It also gave authority for the appointment of officers and the inspection of dairy factories and milksheds on dairy farms. However, the Act came into effect too late for trademarks to be registered so that provision could not be totally enforced in the first season. Nevertheless, the legislation was an important step towards the more important Act of 1894.

The Department's first Annual Report summarised the number of creameries and cheese factories operating in the 1892-93 season, with further details of the value and quantity of butter and cheese production and the scale of factory operations. During the Departments's first season three dairy experts were employed to give demonstrations of butter- and cheese-making: John Sawers as Chief Dairy Expert, with Newman Anderson and WW Crawford as assistant instructors. Crawford resigned at the close of the season but helped set up factories in the South Island. He also conducted a series of lectures and schools in dairy instruction at Banks Peninsula.

In 1892 the Government provided further financial assistance to the Middle Island Dairy Association for experimentation and research into an improved system of ventilation for vessels carrying export butter and cheese. As a result, several steamers were fitted up with the improved system. The Agent-General in London investigated the problem of maintaining uniform temperatures in ships' freezing chambers and a special thermometer was fitted on several steamers.

The Government assisted the industry indirectly by removing the duty on parchment paper used for wrapping butter and also made available free of charge, complete sets of plans and specifications for cheese and butter factories and skimming stations.

Acting upon the suggestion of John Sawers, the first Babcock testers were introduced in 1892. These were hand-powered machines, with an eight-bottle capacity and were manufactured by Burrell and Whitman of Little Falls, Utica in New York State.

In the Annual Report of the 1892-93 season, Sawers referred to the prices realised on the London market — cheese at about £42 per ton and factory made butter at £84 per ton. The Report also refers to the meeting in Dunedin of cheese- and butter-makers, under the auspices of the Middle Island Dairy Association. This was the first of what were later to be termed Dairy Factory Managers' Conferences.

Some indication of the range of skills, breadth of knowledge and technical experience of early dairy instructors can be glimpsed from the details of lectures given on tour during the 1892-93 season: "Co-operative Dairying as an Economic Factor on the Prosperity of New Zealand Agriculture" and "The Dairy Industry and its Relation to Agriculture". These lectures were given in 30 places, around the country.

The Dairy Industry Act 1892 was strictly enforced during the 1893-94 season and a number of convictions were made for neglecting to brand dairy produce properly. Evidence suggests special problems in the packing and shipping of dairy produce.

Shipping conditions for dairy produce at that time were inadequate and unsatisfactory and considerable attention was given to this phase of the early dairy export trade. An important step to improve the shipping situation was taken in 1893 with the Government appointment of Samuel Lowe to report on the condition of dairy produce upon its arrival in London, and upon the efficiency of cool chambers on steamers and various other shipping facilities for perishable cargoes.

In October 1893 CR Valentine was appointed Chief Dairy Expert. His principal contribution to the development of the dairy industry was to lay the foundations for a system of grading dairy produce which was introduced the following year. Sawers continued as a Dairy Expert and JT Lang was appointed as an assistant instructor.

The Department of Agriculture's Annual Report included a "Tabulated Statement, Number of Factories and Creameries and the Extent of their Ramifications", the first occasion on which such information was collected and publicised. Ultimately this information was included in the Dairy Division's annual publication, *The Annual List of Creameries.*

The Dairy Industry Act of 1894 marked the real beginning of progress, for included in it were important provisions relating to the introduction of the dairy produce grading system, the establishment of port coolstores, the appointment of dairy produce graders, the payment for milk on the basis of quality or productive character, the registration of dairy factories and the inclusion and use of such registration numbers in company brands. The Act provided for continuance of the powers of inspection conferred by the 1892 Act. Dairy factories were inspected before they could be awarded special certificates which entitled them to use certain marks. The certificates could be suspended if an inspector reported that neither the dairy nor the factory was kept in an acceptable manner, or if it was thought that the way in which the produce was manufactured was unsatisfactory.

The first four coolstores were established at Auckland, Wellington, Lyttelton and Port Chalmers, with grading officers appointed to each. The first graded produce, a shipment of butter, was received in December 1894. The marking of cheese according to grade did not begin until the 1899-1900 season.

When CR Valentine resigned at the close of the 1893-94 dairy season, his position was filled by JB MacEwan. In May 1895 MacEwan was appointed Chief Dairy Expert and designated Dairy Commissioner. MacEwan resigned in October 1896 and established a very successful merchandising business in dairy appliances and machinery.

Two more dairy instructors were appointed during 1894, SM Robbins and B Wayte. Robbins was from the Mohawk Valley of New York State, with experience both in farm and factory cheddar cheese-making. He had also managed a large experimental factory owned by Burrell and Whitman (later DH Burrell and Company), a world renowned maker of dairy appliances and machinery. In 1885, the Wyndham Dairy Company, to whom Burrells were shipping dairy machinery and plant, asked for a factory manager. Robbins was chosen and arrived in Wellington in October 1885. In the following season he moved to the cheese factory at Gore, where he stayed until his appointment as a Government dairy instructor. During a single season with the Department he planned various dairy factories, some of the most successful southern factories being built and based upon plans, dairy plant and appliances, all of which he supplied himself.

Wayte was born in England, and obtained his early training in both cheese- and butter-making at the Cheshire Dairy Institute. Arriving in New Zealand in 1887, he and his brother formed the Otamita Bridge dairy factory at Otamita in Southland. In 1892 he returned to England and attended the British Farmers' Institute at Aylesbury where he won the diploma and silver medal in 1893. He studied milk analysis under a noted dairy chemist who was also the author of a standard work of the day on cheese-making. Wayte was appointed to the Government dairy staff in 1894 and later assumed responsibility for milk testing sections in dairy schools.

In 1895 the first Government dairy schools were held at Edendale in the South Island and Stratford in the North Island. These short, winter courses lasted about a month during the "dry" season and included lectures and demonstrations of all aspects of dairy factory processing. There were 105 students in attendance at Edendale and 107 at Stratford. JB MacEwan held overall responsibility for the courses.

At Stratford a combined churn and butter worker was introduced and demonstrated. First used by J and R Cuddie at Mosgiel in 1891, it was many years before such a combined churn and butter worker was in common use. The main reason for its introduction was that, in conjunction with refrigeration, it helped to regulate the moisture content of butter. Another special feature of the Stratford dairy school was the Hall refrigerating machine, which was demonstrated there.

In 1896 dairy schools were held at Edendale and Waverley in the winter months, with attendances of 63 and 59 respectively. They were revived in 1900 at Wyndham in Southland, and at the Moa dairy factory at Inglewood in Taranaki, with 43 students enrolled in each. In 1901 the last of the short course, factory based dairy schools for factory managers and their assistants were held at Stratford and Stirling.

For several years it had been the custom for one of the Dairy Division's cheese instructors to hold classes during October and November at different centres on Banks Peninsula. These schools were run for the benefit of cheese-makers in farm dairies rather than those in factories. In the 1901-02 season the available time was too limited for the usual routine to be followed and consequently classes in the final years were organised in three local co-operative cheese factories: German Bay (later Takamatua), Barry's Bay and Wainui. The local premises and equipment were utilised by the Dairy Division staff and private cheese-makers were invited to attend.

In 1895 the Government passed legislation to regulate the manufacture and sale of margarine. It became illegal to process or produce margarine as

New Zealand's first dairy school 1895.

Health Department poster.

an imitation of butter, to mix margarine and butter, or to sell margarine as butter. It can be seen that by the end of 1895, the Government was actively encouraging and supporting the early growth and development of an infant dairy industry. It was also giving considerable financial assistance indirectly by way of services, and directly by means of special grants, freezer charges, dairy schools and the appointment of dairy experts and instructors.

In February 1896, H Grey was appointed by the Agent-General to prepare a report on the condition of New Zealand dairy produce when it arrived in London. The resulting report on the cartage and storage of dairy produce suggested that conditions were far from satisfactory.

Several areas of experiment and research had been investigated during the year. Wayte had carried out some of the earliest testing of cows on a

Waikato farm. Pasteurisation had been widely discussed and supported by Government officals and was introduced at Waverley in 1896-97 the first New Zealand factory to do so. In 1896 the Department appointed two officers as inspectors of dairies because of widespread criticism and concern at the lack of sanitation in so many milking sheds.

In 1898 the most significant developments initiated by the Government were the formation of the Dairy Division as a separate branch within the Department of Agriculture, the appointment of a Canadian, JA Ruddick as Dairy Commissioner and the passing of further dairy legislation, the Dairy Industry Act 1898.

The principal provisions of the 1898 Act related to financial advances to dairy companies by the Government for the purchase of land for dairy factory sites, the erection of suitable buildings, and the purchase of dairy plant and equipment. These sections of the Act were not popular with dairy companies since most were able to obtain suitable finance from banks more expeditiously and with fewer restrictions. On the few occasions when the Government advanced loans it usually lost money. The Act also considerably enlarged the powers of the Department since it removed control of the local milk supply from the local authority in question and placed it in the hands of the Department.

Before 1898, grading applied only to shipments of dairy produce intended for Britain, although Australian dairy purchasers had been asking for two seasons that all dairy exports should bear the Government trademark. In the 1898-99 season the grading system was extended to include all butter and cheese shipped to the Australian colonies. Little of this butter was frozen so to facilitate Government grading special stores were set aside where dairy produce was despatched 24 hours ahead of steamer departure.

The major departmental problems of the season related to shipping facilities and butter and cheese containers. As far as shipping was concerned, the positive impact of the dairy associations is reflected in the 1898-99 Annual Report: "... The facilities for shipping were better.... An agreement has been entered into by the National Dairy Association with both the New Zealand Shipping Company and the Shaw Savill and Albion Shipping Company for a fortnightly service...."

An innovation introduced in 1898-99 in the grading system was the date stamping of butter to indicate the date of manufacture. Protests by British importers led to the dropping of the practice, although it was continued for a few years for butter reserved for local consumption in the off-season. In 1908 date markings by cypher were introduced.

Another Canadian, JA Kinsella, was appointed Dairy Instructor in 1899 and a year later replaced Ruddick as Dairy Commissioner when the latter returned to Canada. In 1901, WM Singleton, another Canadian who was later Director of the Dairy Division, was appointed Dairy Instructor and Grader, an appointment made with the idea of establishing a permanent dairy school. Preliminary plans had been drawn up and land accepted in Palmerston North, but controversy arose over the location and form of the school and, failing agreement, the project was abandoned.

From October 1900, a grading system similar to that for butter was introduced and applied to cheese. Before this date cheese had not been stamped according to grade, although an official examination had been made and the manufacturer supplied with a copy of the grader's report.

12

MILK SUPPLY

When dairy cows were introduced to New Zealand most settlers owned only one or two, thus there was little, if any, special provision for milking. The cow or cows had to be rounded up, often a major exercise in itself as stock were left free to wander, grazing and browsing in the nearby bush and roaming as they pleased over wide areas in search of feed.

The traditional practice was to hand milk cows in a nearby bush clearing, out in the fields or in a yard close by the farmhouse or dairy, with bucket and small stool carried to the cow. Milking conditions were far from easy — the yard was inevitably fouled and muddy, cows had to be rounded up over long distances in rough country, milking buckets and stools were unstable, utensils were difficult to keep clean and milking hygiene was almost unknown.

As fixed milking locations and sites became more common, cow bails were introduced to hold the cow still by constraining its head, but hygiene and cleanliness problems persisted with the muddy conditions. The bails consisted of two upright, wooden posts, placed in the ground and with the cow's head locked firmly in between, a smaller wooden bar was placed over the head to join the posts and hold the cow still during milking. The animal's rear leg on the side of the milker was also roped across the other leg to hold it steady and prevent kicking. As far as was practicably possible cows were generally milked by the same milker and at the same time each day to create a familiar routine.

Milking the cows was generally the responsibility of the farmer's wife and children, since this freed the principal breadwinner to carry out the more important and also the more physically demanding tasks around the farm. The yard itself, the scene of various other farm activities, was seldom the cleanest or most convenient place for milking. Milking was a difficult and time-consuming task that had to be performed every morning and evening regardless of unfavourable weather conditions. Even among some of the more religious communities, where factories were closed on Sundays so as not to receive supply and where milk consequently remained on the farm for butter- or cheese-making, milking was still carried on regardless of the sabbath. The only respite in the daily milking grind came when the cows were eventually "dry" at the end of the season.

Labour on dairy farms had always been scarce and therefore expensive. Consequently efforts were made to minimise labour inputs and to

A weary farmer's wife awaits
her husband's return.

maximise levels of productivity with existing labour by the adoption of
mechanisation. The first innovation in this direction was the introduction
of the herring-bone milkshed. The most recent development to be widely
used by farmers with large herds is the rotary milkshed. These enable
farmers to milk large numbers of cows either single-handedly or with
minimal assistance. Mechanisation of milking methods and subsequent
rural electrification and reticulation were certainly prerequisites for the
increasing scale and size of dairy farm operations.

Parallelling the quest for increased productivity, was the pressing
demand for higher levels of hygiene and cleanliness in the handling and
subsequent processing of whole milk. Clean milk made quality control in
dairy processing possible and was a major factor in the adoption of the
factory system. Later efforts to improve both the quantity and the quality
of milk output included the introduction of pasteurisation of cream and
whole milk for processing as well as the adoption nationwide, of butterfat
testing and herd testing.

Improvements in mechanical milking had social benefits too, nowhere
more so than in the reduced dependence on the labour of the dairy
farmer's family, for regular morning and evening milking. Many early

"White slaves of the dairy industry" photographed at the express request of the Premier.

photos taken on dairy farms clearly illustrate that milking was essentially a family affair. The earlier responsibility of the wife for butter- and cheese-making was such that the farm dairy had frequently been thought of as a mere adjunct of the kitchen. Any cash or other return from surplus butter and cheese sales was viewed as pin-money to help with kitchen provisions or family needs.

Such demanding work inevitably took a social toll upon the lives of younger farm children, particularly in terms of opportunities for schooling and recreation. Some city newspapers even went so far as to comment in their headlines on the "sweated labour of so many women and children" found on small dairy farms. The alternative viewpoint was, of course, that the labour of women and children was vital for the very survival, let alone expansion, of such farms and the farming system in general.

The task of milking as many as five to ten cows each morning and then facing a walk of up to 8-10 kilometres to the nearest sole-charge school resulted in large numbers of country children from small dairy farms being too tired to learn at school, probably attending only for a chance to rest from the daily drudgery of looking after dairy cows. After walking home at the end of the school day the same milking routine began again with cows to be seen to before an evening meal and bed. There was little, if any, time for play. Many teachers were sympathetic to the needs of such children for sleep and rest during school hours and made as few demands as possible on them. Reports suggested that this practice of school children milking cows was widespread in the developing dairy districts. Some even referred to the stench of children's clothing which, having

An early sole charge bush
school.

been worn continually in the milkshed in all weathers, was most
unpleasant in the warm classroom conditions. When a Government
Commission was set up to investigate sweated labour conditions among
women and children employed in factories, one national newspaper
suggested that it was not really necessary to look any further than the
employment of children on dairy farms in what it called "the country's
own special white slave traffic".

The early reluctance of many senior Government department officials
to encourage the wider adoption of home separators was because farm
milking was generally carried out in conditions that were totally
unhygienic. They did not therefore wish to perpetuate or expand on-farm
handling or processing of milk. It had already been recognised in official
quarters that the consistent production of high quality dairy products was
possible only with complete quality control. This in turn was impossible,
even with factory processing, unless the whole milk supply was of the
highest quality and handled throughout under the most hygienic conditions.

**Above: Milking time,
Golden Vale, Otakia 1900.**

Right: At work in the dairy.

Milksheds

One of the first efforts to improve the handling of milk on the farms was to raise the standard of hygiene in both the milking process and in the milkshed. While milking in open yards and clearings was unsatisfactory the earliest milksheds were little better, since they lacked concrete floors, adequate drainage, a supply of fresh water, and water-heating facilities. Departmental inspection began with local town supply dairy premises, the first move to extend such inspection being made in 1908. Inspection was shortlived, however, as it met with serious opposition, but it ultimately eventuated in a scheme whereby dairy farm instructors, in co-operation with the dairy companies, supervised the cleanliness of milksheds of factory suppliers.

An official report for the Department of Agriculture in relation to the Auckland district provides some idea of the conditions of dairy sheds nationwide at the turn of the century:

> ... the sheds as a rule are built entirely of wood, the majority having iron roofs. Most of them are in a fair state of repair .... Some half-dozen dairies possess cowsheds which are nearly perfect, ... having floors of either concrete or brick, so laid that drainage waste flows away and is taken in underground pipes to a sufficient distance from the building. Most sheds, however, fall far below this standard. Few are so situated as to ensure and secure good drainage, ...

Waikato Winter Show, circa 1910.

The floors of the milksheds in terms of materials and design posed various problems:

> ... Many floors are of large stones, very roughly laid or simply of earth .... Another form of floor, even more objectionable than the last-mentioned, is the floor formed of earth or large stones to a depth of three feet, with boards laid above. The excreta soaks into the boards and penetrates beneath them, and it is only necessary to stand on the boarding to prove this, as the weight of the body pressing down on the boards forces the filthy fluid to the surface....

In 1900 the milksheds for factory supply were as primitive as those described above. In view of this it is not surprising that officers of the Dairy Division, and others responsible for the manufacture of butter and cheese, kept up constant pressure for improvements in the design and operation of dairy sheds. The condition was not remedied, however, until the Dairy Division issued plans for standard walk-through milksheds.

The real need for improvement in the conditions of milksheds was recognised in the Dairy Industry Act 1908. The regulation of shed inspection did not eventuate, but a great deal of valuable work was done by inspectors without using statutory authority.

Progress in improving milksheds was greatly facilitated by rising dairy prices, while improved roads made it easier to cart shingle and put down concrete yards and floors. In short, farmers made rapid strides in improving conditions as soon as their circumstances enabled them to do so.

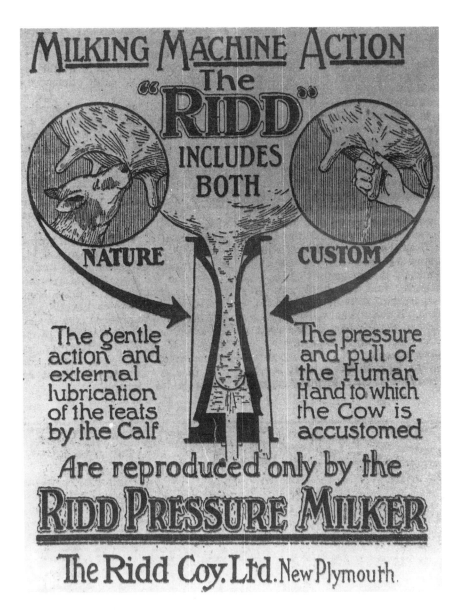

The Dairy Division of the Department of Agriculture not only made available standard plans of well-designed, walk-through milksheds, it also provided instruction on the on-farm handling of milk and cream. By the 1920s these factors had resulted in massive improvement in the standards of hygiene and cleanliness in milksheds and in the handling of milk and cream on the farm.

Mechanical Milking

Along with improvements in milksheds there were comparable advances in mechanical milking techniques. The concept of milking cows by using machinery was not a new one and even before 1894 various kinds of milking machines had been patented. At least two models, the "Nicholson and Gray" and the "Marchland's Patent Milking-Machine" were already on the market. In 1894 R Withell of Brookside in North Canterbury patented the "Withell's Brookside Patent Mechanical-Milker", which was manufactured locally in Christchurch. This machine was similar to many of its predecessors, although technical improvements made it much more efficient in operation. The Department of Agriculture was interested enough to have the innovation checked out and it was reported on favourably. Although the patent was never widely taken up it was the first in New Zealand of sufficient merit and promise to demonstrate the real possibilities of mechanical milking. Officials were not agreed on its desirability, however. Some openly opposed the introduction of any form of mechanical aid for milking. Others, while appreciating that much of the inferior milk supplied to factories may have been the result of the lack of proper hygiene in the cleaning and maintenance of milking machines, also recognised their potential for

Early mechanical milking, Taranaki, circa 1912.

Below: Herringbone milk shed, Ruakura.

future industry expansion.

The early development and adoption of various brands of milking machines on smaller dairy farms, particularly in the Auckland and Taranaki districts in 1900, meant that farmers' children had to be called upon less and less to assist with the added demands for milking larger numbers of cows.

Although milking machines had been used before 1900, the Lawrence-Kennedy machine, invented around 1902, was the first to provide a satisfactory milking unit, and gain a measure of general acceptance. These machines were imported from Australia in 1903, one machine imported that year being tried out at Ruakura. They were followed by the Wallace in 1907 and the Gane Releaser in 1908.

Early opinion favoured mechanical milking and the Department of Agriculture's Annual Report of 1904, recorded the milking machine as an unqualified success. In particular it facilitated the increase of dairy cow numbers and the enlargement of herd sizes, while raising levels of labour productivity and consequently the wages of dairy labour.

The Dairy Division, however, claimed that because of the practical difficulties of cleaning the machines thoroughly, the quality of milk from the earlier, machine milked cows was generally unsatisfactory. Gradually, the design of machines was steadily improved so that all parts could be

The herd returns from evening milking.

readily taken down and cleaned, while the routine washing and care of dairy machines was carefully systematised.

The main concern of dairymen in relation to the new milking machines was the fear that there might be damage to the cows' teats or interference with their milking quality.

By the 1920s the efficiency of the milking machines available had induced a general changeover from hand to machine milking. With dairying expanding so rapidly the demand for adequate labour for milking was such as to accelerate the wider acceptance of milking machines in normal farm practice. The main hindrance to expanded use of the machines was the initial lack of suitably cheap, reliable motive power and the cost and poor quality of rubberware. Wartime labour shortages hastened developments so that by 1918-19 7,577 machines were installed. As a result nearly 50% of all dairy cows were machine milked, the percentage increasing to 86 by 1941. In the intensive dairying districts of Auckland and Taranaki, the number of cows machine milked was 95% and 93% respectively.

13

PRODUCTION SYSTEMS

From the early years of the 20th century the home separation system spread until finally, some 20 or 30 years later, it completely superseded the earlier creamery system of dairy processing and production.

Home Separation: The Pros and Cons of Rival and Competing Systems

Home separation for farmhouse butter-making had been a fact of life in some districts since farm separation had first been introduced in the late 1880s and early 1890s. This was particularly so in those districts where whole milk or cream deliveries were impracticable because of isolation or rough terrain. Farm separation, or home separation as it was more commonly termed, had a long and chequered history in the early decades of the 20th century, with prolonged debate over the pros and cons of the new system. It was not until the 1930s that it became generally accepted. The debate was one which engendered tempers and feelings as never before in the history of dairying. Protagonists argued hotly about the merits or demerits of the rival systems — the existing central creamery or the new innovative home separation system. Industry leaders, like Wesley Spragg and William Goodfellow were sharply divided on the relative advantages and disadvantages of the two systems, as were suppliers and managers. It was not until the late 1920s, when more than 90% of all milk supply was processed by home separation methods, that the issue was more or less resolved.

The changeover to home separation and factory cream delivery was gradual for although many companies supported the new system, suppliers transferred only one group or area at a time. A glimpse at the returns set out in dairy company annual balance sheets of the 1920s and 1930s, reveals the extent and rate of the changeover in individual companies. The continuing dominance of the whole milk supply reflected the persistence of the creamery system, with suppliers delivering whole milk for separation in the central creamery or in one of the associated skimming stations. However, an increasing supply of cream delivered directly to the central creamery, either by company or contract can collection, was indicative of the gradual spread of the new system. In many companies the changeover took as long as 10 to 15 years. As the balance of supply shifted steadily in favour of cream rather than whole milk, the home separation system gained wider and ultimately, total,

Delivering cream to the factory.

industry acceptance.

Interestingly, strong opposition to the new system came mainly from the staff of the Department of Agriculture's relatively new Dairy Division, butter graders, farm and factory inspectors, and from various influential dairy factory managers, dairy proprietors, and other industry leaders.

Dairy Division opposition arose from concern with the standards of hygiene for handling milk on the farm. Poor quality milk or cream inevitably led to poor quality butter. When the factory system of dairy production was originally introduced for export butter-making, the most widely held opinion was that, in the long-term, the interests of the dairy industry were best served by using power-driven separation at the manufacturing dairies or in the associated skimming stations. This was particularly so in order to produce and maintain the highest possible quality for export produce. Inevitably it meant daily delivery of whole milk to the dairy factory, immediate separation and cooling, followed by the ripening of the cream and the addition of a mildly soured and separated skim milk, the forerunner of the later "starter" cultures.

The first deviation from the central factory system had been the introduction of networks of skimming stations, the so-called creamery system. It was a system which had originated in the dairying areas of the United States, in the states of New York and Wisconsin, where it was often referred to as the "associated dairy system". It had spread to New Zealand, along with other American dairy expertise and experience, practices, plant, and equipment.

The new system spread rapidly because no strong or convincing case was ever made against its extension. Occasionally there were requests from some dairy farmers for special dispensation to deliver farm separated cream to the factory for manufacture. These requests usually

Transporting milk cans in the appalling winter conditions.

came from farmers in isolated or rugged hill areas, who because of their geographical location were unable to deliver whole milk to creameries or skimming stations. In most cases the reply they received was to make their cream into farm butter for sale, and this they duly did.

Although by the turn of the century the creamery system was firmly established in New Zealand and whole milk delivery was considered to be in the industry's best interests, in the United States and Australia home separation was becoming widely accepted. The particular argument advanced in favour of the creamery system was that the whole milk was under the close scrutiny and control of the recipient factory through all stages of manufacture. Moreover, factories were able to accept or refuse milk supply, as well as advise the supplier of its condition and on any need for and means of improvement.

For many years after the export dairy trade was established, one of the most serious and frequent faults found with the quality of export butter was its decidedly "fishy" flavour. It was generally accepted that the departure from the central factory system had been at least partly responsible for this problem. This was because much of the home separated cream was older and consequently more acidic than central factory cream, a direct result of infrequent deliveries and undesirable on-farm handling conditions. It was an accepted premise that fresh, sweet cream was the most important prerequisite for high quality butter, and

any departure from such basic principles was scarcely prudent, and must inevitably be detrimental to quality.

Many who staunchly supported the creamery system also admitted that there were economic advantages in the home separation system. The most serious limitation of the new system was that the manufacturing dairy had little control over the quality or handling of the whole milk before delivery. Such limitation presented serious disadvantages as far as the quality of the resultant butter was concerned. When home separation was finally established there was an immediate and noticeable increase in fishy or tallowy flavoured butter and, needless to say, the ranks of the critics and opponents of home separation expanded rapidly and significantly in response.

By 1905 home separation was already on the rise, just as the creamery system had reached its peak. The only direction the latter could take was downward. Another matter of historical interest was that, at about the same time, traditional farmhouse butter- and cheese-making had been almost totally superseded by factory production. There were two relatively isolated exceptions, Banks Peninsula, where farm cheese-making survived, and Taranaki, north of Waitara, where farmhouse butter-making persisted.

Despite those critics who deplored its growth, the home separation system spread. In 1905 two companies, Whenuakura and Mangatoki, co-operated with officers of the Dairy Division to pasteurise whole milk cream as a safeguard against fishy flavours. Then in 1907, some butter makers handling farm separated cream for manufacture used a neutralising agent to reduce levels of acidity. When the addition seemed to prove useful it came into general use.

The introduction of home separation in New Zealand can be traced back to 1897 when the system was introduced by George Finn of Messrs Finn, Chisholm and Company of Wellington. The firm, having obtained the agency for the Sharples farm separator, had successfully persuaded some dairy farmers of the efficiency of the new machine, as well as of the benefits of on-farm cream separation. One particular advantage was the much smaller bulk of raw material that had to be delivered to the factory, almost 90% less than the quantity of whole milk that was required. In those days of extremely bad roads, milk delivery could often prove very difficult so that distance from the creamery or skimming station was an important factor. It was often impossible and involved great risk and difficulty, to use a wheeled vehicle of any description, and there were even some back-country roads, or perhaps more accurately, tracks, which at certain times of the year could not even be negotitated by a wheel-less sledge. Consequently suppliers frequently had to resort to pack-horses to deliver their milk. It was in isolated locations such as these that dairy farmers welcomed any means of lessening the time and labour involved in delivering large quantities of milk to the factory for processing. Home separation enabled just such savings.

Although Finn, Chisholm and Co was successful in persuading farmers of the more obvious advantages of the new system, the company soon discovered that it was a much more difficult task to persuade factory management to accept a supply of farm separated cream for processing. Confronted by this dilemma the company, probably initially as a sales gimmick, built a dairy factory of its own to accept and process farm separated milk. The dairy business operated for about 8 years both as farm separator agents and as manufacturers and distributors of "Gold-leaf" butter which was made from farm separated cream. Later, the business developed into one of the most well-known concerns in the city,

Opposite page above:
Auckland Agricultural Show, Alexandra Park, Epsom 14 November 1902.

Opposite page below:
Okato cheese/butter factory circa 1910.

the Wellington Fresh Food and Ice Company.

In 1903 Joseph Nathan and Company introduced, or rather accepted, home separated cream at their "Defiance" factory at Makino, near Feilding, and was probably only the second company to do so.

It is significant that a dairy appliance catalogue issued in 1883 by JH Monrad of Karere at Longburn, near Palmerston North included an advertisement for hand separators, in effect, foreshadowing the later home separation system. Probably the first illustrated catalogue of its kind, its coverage was comprehensive including both cheese- and butter-making utensils and appliances. As well as outlining and describing the current creamery system, Monrad also anticipated the home separation system and the lines upon which it would operate. He also pointed out the need for mechanical refrigeration in factories to control temperatures in the manufacture and storage of butter — a prophetic comment for 1883!

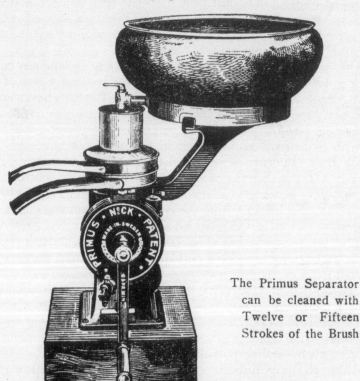

One of the many separators available to the farmer.

The Home Separation System: Its Influences and Consequences

By the 1930s and 1940s the home separation system was well and truly developed and firmly established. It is interesting to observe the effect it had upon manufacture, which included the establishing and equipping of dairy factories and to note its impact on the day-to-day life and livelihood of dairy communities.

The introduction of the hand separator facilitated the skimming of milk on the farm. Accessibility to the factory became a critical consideration for suppliers and an almost determining factor for a successful dairy farm. In earlier years, milk suppliers who lived in areas with bad roads and difficult access were, if necessary, given special dispensation in the spring months to separate on the farm and deliver cream until the impassable roads of winter had been dried by the summer sun. Having minimal knowledge of cream testing, factories measured rather than weighed samples, which meant they were paying for more fat than they actually received. This may have resulted in the popular notion that home separation and delivery of cream gave the farmer a better return than delivering whole milk.

The new home separation system was gradually extended to more existing suppliers, and also to newly developing dairying districts where it was found most suitable. Districts with fewer cattle and consequently, a comparatively small milk supply could now support a factory because the supply area, or milkshed, extended over a much wider area. There was no longer the added expense of establishing and equipping skimming stations. Moreover, the saving in daily delivery time gave the supplier more time for developmental work on the farm, and the plant required for on-farm separation probably cost less than that for milk delivery.

The development of the power-driven milking machine led to a

Mrs Johnston's dairy, Manurewa.

comparable use of power attachments for home separation, especially where larger herds were involved.

One of the important results of home separation was the final elimination of the export trade in blended farm butter which had been produced in packing stations. Farm butter-makers now found a preferable alternative which was certainly more profitable and labour-saving. This was to sell cream rather than put so much energy and effort into butter-making.

Initially the farmer delivered cream to the factory. However, this system was soon superseded by community delivery. Company or contractor-based collection and delivery of cream was organised by the dairy companies and was usually paid for on a per trip basis. The costs were charged against general expenses and all suppliers were debited at the same rate regardless of the distance involved. This costing frequently determined the general location and more specifically, the siting, of the dairy factory because all suppliers were paid the same price for fat. Later the companies used their own vehicles and driver to collect the cream.

In summary, the system produced several results. It provided suppliers with a choice of alternative factories to supply, not necessarily all in their immediate area. At the same time it allowed dairy companies to extend their own supply and collection areas further afield. By enabling dairy companies to expand the areas from which they drew milk supply, the average output per factory was significantly increased. New Zealand factories averaged a larger production output, on a per factory basis, than many of their Danish counterparts.

The rationalisation of processing and production units resulted in the closure of several dairy factories. In fact, after the first peak in 1905 when the total number of dairy processing units was 1,063, there was a steady decline until in 1935 there were only 542, a total which included butter and cheese factories, skimming stations, packing stations and registered private dairies. Amalgamation and absorption facilitated the settlement and development of new land and home separation made smallholder dairying, then fairly widespread, both a viable and practicable enterprise. An indication of this was the large number of small suppliers producing a given quantity of fat at home separation factories compared, for instance, with the also traditional small cheese factories. It facilitated the spread of small suppliers as well as the establishment of small dairy factories in some of the relatively isolated districts.

On the other hand home separation had less welcome effects. For instance, it resulted in the loss of the earlier close personal contact between factory managers and suppliers, and between the suppliers themselves as they queued daily at the skimming station or creamery to await delivery of their milk or to collect skim. This regular, close and dynamic contact was never again completely realised.

There were practical problems too. The small quantities of cream supplied in spring and autumn resulted in less frequent deliveries. This was particularly unfortunate since it led to a fall in the quality of butter manufactured at those times. It was impossible to weigh the thick, sour cream effectively since it clung to the sides of the factory weigh cans. Consequently a system was adopted whereby suppliers' cans were weighed individually using tare weights. Contractor collection and delivery of cream cans also meant suppliers had to label both cans and lids. Cream cans had to be washed before they were returned to the supplier and this led to a demand for extra factory space in order to handle the large numbers of cans involved. Ultimately this resulted in the

Motorised collection of milk.

introduction of mechanical conveyor belts and automatic washers to ensure quicker handling and turnaround. Later it became necessary for factories to provide for the daily testing of cream, and later still cream grading was adopted to cater for the variable and frequently inferior supply brought about by home separation.

In terms of factory construction, the elevated receiving platform with its hoist to raise milk cans had to be lowered to a more convenient level for vehicles. Eventually a level factory floor was adopted in place of the gravitation or gravity floor system, and cream pumps and pressure systems were introduced.

As far as manufacturing was concerned, the sour cream which was received caused cooling difficulties and resulted in sour, inferior butter. Pasteurisation was introduced to offset this effect and to improve quality. Those dairying districts which maintained daily deliveries of a good class of cream made better butter with the help of the pasteuriser than that which was made under the whole milk system without the use of

Feeding the pigs with skim.

pasteurisation. Further the butter had a much fuller and cleaner flavour than that made from neutralised cream.

By the late 1930s, larger herds reduced the distance that had to be travelled to procure a full load of cream cans, and improved roads made motorised transport practicable. Both factors led to a reversion to daily collection over much of the dairying area of the North Island and, consequently, butter of better quality.

There were other difficulties. Competition between dairy factories and overlapping supply areas led to significant increases in collection costs, difficulty in maintaining a common standard for grading milk and cream at both butter and cheese factories, and lack of a regular and reliable supply, especially to cheese factories. This in turn led to concern because of the expenses incurred in maintaining adequate supply and efficient equipment. As a result, dual plant factories were developed to retain supply, and the admission of suppliers without shares was increased to boost supply. This was a real deviation from the basic ideals of co-operative dairying.

Finally, the rationalisation in both supply and production facilitated by home separation, resulted in a move away from the original idea of a large

number of smaller creameries and skimming stations owned by one company, and operated in conjunction with a single main factory. Increasingly, the tendency was towards centralised control and large factories which were functionally complete in themselves, part of the trend which led the supplier to be further removed from basic decision-making.

Home separation certainly assisted in the settlement, breaking in, and development of lands in new dairying districts, but its adoption initially raised some very real misgivings. In the early years of the system's operation some districts did not insist on a daily delivery of cream and as a result the butter manufactured was not always of the highest quality. However, with a regular delivery of fresh cream there was a marked improvement in the quality and output of butter.

The facilities provided by dairy companies for delivery of cream to central dumps, where it was weighed, sampled and emptied into large cans for transport in bulk to a central creamery, eventually led to much overlapping of supply areas and to cream can haulage over long distances. This proved costly and involved funds which could have been better utilised. Also, some of the needless competition which resulted led to the closure of many of the smaller cheese factories. Sometimes the voluntary sale of smaller dairy factories to co-operative dairy companies was organised, such sales included payment of compensation to the previous owners on the basis of the amount of butterfat received. On other occasions amalgamation proved to be the best course of action. Some "dumps" were closed while for others a transfer or exchange of supply was arranged. Compulsory zoning of supply or suppliers to a factory was not envisaged unless the company to receive the supply of butterfat was able to pay a price comparable to that of other companies. By these means dairy companies hoped to reduce the costs of cream collection and manufacturing in creameries where a large output was assured, and at the same time, to improve the quality of the butter manufactured.

Home separation therefore, had varied results. Spared of the time and effort needed to cart milk to the creamery or skimming station, dairy farmers were now able to devote more attention to final clearing and developmental farm work and to expanding pasture and winter fodder crop areas so they could carry larger number of dairy cows. Dairy production rose rapidly, particularly when rural electrification made power more readily available to operate milking machines on farms.

It was not until after 1910, and in most areas after 1920, that home separation had really taken over. Skimming stations in South Taranaki became redundant and either closed or switched to cheese production. At the same time, with the development of the internal combustion engine, reliable motor transportation spread, drastically and abruptly changing the economics of factory supply. Company or contract lorries and trucks carried cream much longer distances, significantly extending factory supply areas with less time and effort and lower costs. Dairy output expanded as did the number of cows and suppliers. Many dairy factory operations were amalgamated, resulting in the dominance of a much smaller number of larger factories.

CHAPTER

14

SCIENCE AND TECHNOLOGY

In colonial New Zealand the first settlers had encountered a physical environment which was totally alien to them. Consequently, adjustment and adaptation, along with invention and innovation, were the underlying rules for survival in day-to-day living, especially in farming activities. Not all settlers were agreed on the need for change and an early conflict implicit in the growth and spread of European style farming arose between those who favoured the replication of the more familiar styles of British farming, and those who appreciated the need for change in the approaches and attitudes demanded by the new bush and scrub environment.

Pioneer Grassland Scientists: Their Principal Contributions

In the early decades of the 20th century a new form of colonial, all pasture dairy farming emerged. This was mainly based upon a favourable grass growing climate, but even more upon the research of a group of grassland scientists who systematised the method providing it with a truly scientific basis.

R G Stapledon

Stapledon's research in Britain indirectly related to the New Zealand grassland in that he raised common concerns by revealing the deplorable botanical state of grasslands in Britain. Most existing British pastures were shortlived because of the use of poor pasture strains, and had evolved largely through a constantly repeated system of sowing and reaping for seed under a predominantly arable farming system. Little, if any, attention was given to pasture selection and breeding and the system was limited to those strains suited to quick establishment of annual pasture species and plants.

Similar deterioration was fast taking place in New Zealand pastures. The widespread and common practice was to save seed from South Island, short-rotation farming systems. Shortlived, more or less annual species of ryegrass and white and red clover were being evolved but these failed to establish suitable permanent or long rotation pastures.

Stapledon set out to improve the quality of British grasslands. He set up a plant breeding station with a very strong grassland bias and although he made some use of imported varieties he was strongly in favour of a "home grown seeds are best" approach. This pattern was also followed in New Zealand although it was slightly modified by the importation of some foreign, high producing pasture species from other temperate regions.

Stapledon's visit to this country in 1926 paid off well. It resulted in the setting up of herbage strain testing in 1928, and ultimately in the certification of high quality grass seed for export as well as for the renewal of local pastures.

Stapledon's real concern was demonstrated by his insistence upon the significance of strain and breeding in the improvement of grasslands. His brilliant understudy, Davies, was seconded to New Zealand for 2 years to assist with strain testing in order to establish the importance of strain and breeding in pasture improvement.

AH Cockayne

It has been claimed that Cockayne was the founding father of scientific thought on grassland farming in New Zealand. He was an eminent grassland philosopher but pragmatic rather than idealistic. He adopted an ecological outlook in respect to pasture associations, viewing them as dynamic ecological entities, readily moulded and modified by changing environmental conditions. This perspective is evident in much of his earlier writing about the conversion of scrub, fern, forest and bush to grass. He repeatedly emphasised the profound and far-reaching importance of research based upon ecological concepts. The tussock soon attracted his attention producing a critical assessment "that the system then practised of farming the country by regularly burning the tussock would upset the delicate balance between the vegetation and the environment".

In 1910 Cockayne stressed the overall importance of the New Zealand grass crop which overshadowed all others. Like Stapledon, he envisaged the setting up of a plant breeding station to improve indigenous pasture plants. He was a strong advocate of the use of clean and verifiable seed, and of a sound seed control system. He urged the establishment of a plant breeding station for the production of improved, botanically sound pure grass seed types in order to produce the necessary "pedigree" and registered type of seed for testing. Ultimately all seed sold would include its pedigree tag or certification and guarantee as to type, purity and generative capacity.

In 1930 Cockayne, arguing for the recognition of the significance of "strain" as a factor in grassland development, claimed ". . . of more real importance, and at present barely recognised, is the question of strain — the pedigree grass capable of maximum returns under the hardest stocking conditions that obtain under rotational grazing — and full utilisation of all feed produced, is really as important as pedigree livestock. Recognition of this fact is likely to play an even more important role in the improvement and development of New Zealand grasslands than any other milestone that has been passed in the history of New Zealand farming".

He was, from early times, an ardent advocate of topdressing for agricultural development. Applied to all types of land, especially developed and developing grasslands, this was a means of successfully controlling secondary growth of all kinds. Cockayne's so called "veneer of mineral fertiliser" during the years 1920-1930 played an important role. He claimed the paramount factor in grassland development was the increasing part played by topdressing. It provided nitrogen "factories" by regularly fuelling the clovers with phosphates, the most potent factor in bringing about high per acre productivity.

In 1936 Cockayne further recognised that "... aerial topdressing is increasing at a great rate, an increase from none to 400,000 tons in 5

years.... It makes possible the topdressing of all grasslands in New Zealand, impossible to achieve by land machinery or by hand".

Reaping and gathering grass seed, Pohanga Valley, circa 1890.

The New Zealand Grasslands Association was formed in 1931 and at its first conference in 1933 Cockayne commented that because "... a rain-forest climate is synonymous with high production grassland potential, an intensive type of grassland farming has evolved, having for its objective the production and utilisation of milk producing pasture for the cow, ewe or sow — the essential elaborating machinery for our grass crop into butterfat and rapidly maturing meat. There are great potentialities of New Zealand increasing her grass crop as well as improving the utilisation of this great crop ...".

In 1939 Cockayne, now Assistant Director-General of Agriculture, outlined the management methods needed to make the utmost use of the female animals. These were to obtain the pasture composition and growth pattern suitable for wet stock, to lengthen the period of pasture growth, and finally, to increase the production of high protein grass demanded by wet stock through the greatly extended practice of topdressing.

PD Sears

A background in geography did not stop Sears from developing into yet another plant ecologist. He made a particular study of those factors

139

governing the establishment, growth and development of pasture species and pasture associations, and the appropriate livestock associations. Sears took a practical approach, using as his laboratory both the field and the farm with small experimental plots and farmlets, and conducting trials on a sound statistical basis. The most important and far-reaching outcome of Sear's work was the increasing significance attributed to the extremely close relationship between the grass sward and the available grazing, as well as the vital roles of clovers and the management of the grazing animal. Research was developed to determine and analyse the composition of the feed intake and its utilisation by the grazing animal, as well as residues returned to the soil or pasture as urine or excreta. This ultimately resulted in the formulation of a fairly simple yet pertinent grazing philosophy: stock need grass just as grass need stock, or alternatively, the greater the number of livestock grazed and adequately maintained on any pasture, the more fully and effectively the available pasture will be utilised and, over time, the more productive it will become.

Sears emphasised the need for a high level of clover production to maintain pasture productivity and carrying capacity. He advocated the application and wider use of fertilisers in the initial stages of pasture development, as well as the adoption of appropriate grazing practices and management policies to ensure high stocking per acre capacity. As a result of the latter, soils received large inputs of excreta and urine which returned the essential nitrogen for grass and pasture renewal. Sears was therefore largely responsible for introducing the concept of a highly intensive, all pasture system of grassland dairying. Sears also highlighted the substantial losses in dry matter and subsequent livestock feed and nutritional value that occurred when pasture was conserved for making hay or silage or when it was trampled by stock.

E Bruce Levy

Levy also adopted an ecological perspective. He was especially concerned with the advantages provided by New Zealand's climate, particularly in the areas of pasture growth and the length of the grass growing season. These were much better than in the Northern Hemisphere.

Like Sears, Levy appreciated the role of the grazing or browsing animal in maintaining high levels of productivity in pasture. He was keenly aware of the potential in New Zealand for further expansion and intensification. Writing in the mid 1930s, Levy claimed that evidence suggested much of the grassland lowlands, then currently yielding less than 200 lbs of butterfat per acre (or 3-4 ewes plus cattle equivalent) were potentially capable of yielding close to 400 lbs of butterfat per acre.

Levy endorsed Sears' comments on higher stocking rates as a basic requirement for continuing pasture development and improvement, adding further that "... only with stocking rates of 1 dairy cow per acre, or 6 - 8 ewes plus cattle, will the grazing system return fertility to the land to provide a supply of plant food for the grass crop". The critical role of both clovers and minerals was stressed, the latter providing for bulky and healthy clover growth. He noted that apart from phosphates, lime, potash and molybdenum, the most important and vital element of all was nitrogen, which was manufactured in the root nodules or "factories" of clover. This was then elaborated by grazing animals into animal protein or dung and urine wastes. These wastes put further nitrogen and minerals back into the soil enabling the building up of the grass sward and the continuation of the fertility cycle. There was therefore little point in

growing grass and producing pastures unless the feed was effectively grazed in situ and residues or wastes passed back to the sward.

The aim of both Levy and Sears was to produce more intensively, per acre, irrespective of capital or labour input. It was a matter of fully optimising the climatic advantage; New Zealand's climate provided ideal conditions for the growth of lush, high producing, high quality grass and fodder crops, as well as year round outdoor grazing.

Traditional Theory: The Annual Versus the Perennial

Agriculture the world over has traditionally been dominated by the "annual", especially where there are critical climatic periods during the yearly cycle of production. In the great wheat belts of the world the annual is fully exploited, because it can establish itself at the onset of spring when conditions become suitable and grow rapidly and ripen its seed before cold or drought makes growth unsatisfactory or impossible. Thus the annual can survive between critical growth periods as a cold and drought resistant seed either in the ground or the granary.

If these same areas had to rely on perennials, such perennials would have to be low yielding, tough leaved, deep rooting, frost resistant, and winter dormant plants that cease to grow during early autumn and begin to grow again late in spring. The only traditional perennial grasses for such conditions are those with underground stems and crowns.

Extreme continental conditions therefore make it impossible for perennial pasture plants to play an important role in agriculture. Very cold winters and hot summers favour cereals and other bulky annuals which establish, grow and mature quickly. The only perennials capable of tolerating such climatic extremes are low producing, hardy, and for the most part underground creeping species. For example, in the United States Kentucky Bluegrass is capable of tolerating extreme winter cold and surviving summer heat and drought. However its yield is only about one-sixth of a good ryegrass and white clover pasture in New Zealand.

In New Zealand the perennial ryegrass strain which is so widely used is fairly resistant to both cold and drought and survives summer heat better than white clover, its partner in the pasture association. Not only dairying, but also fat lamb production in New Zealand depends largely upon perennial ryegrass for winter and early spring feed. Perennial ryegrass was therefore the major outcome of the "Grasslands Revolution".

Evolution of Early Grassland Farming in New Zealand

The most important direction taken in grassland management was to minimise the seasonal nature of grassland production. Increasing use was made of female animals such as the dairy cow, which had a dry period during winter. The proportion of dairy cows to total cattle wintered steadily increased, a trend in dairy development from which the two outstanding theories of grassland management evolved. These were, first, that the feed requirement of the grassland farmer should be at its lowest in winter, and second, that summer grass production should be converted into milk represented by the returns the farmer received for butterfat.

The latter theory meant it was essential that herbage should be of a milk producing type. It must be young, grow vigorously, have a high mineral content and available protein. Good quality grass in New Zealand contained up to 30% protein making it remarkably good for milk production.

In the 1930s New Zealand grassland farming practices were centred on the development of three phases: improved conditions for pasture growth; improved management and facilities for better utilisation of pasture growth; and use of stock which could use the improved conditions more efficiently.

The distinctive features of these phases were drainage, fertilising, liming, surface cultivation, hay and silage making, mowing of surplus growth of grass, smaller paddocks, better watering, shelter and disease control, and breeding under tested control. The most significant single factor, however, was topdressing. The area topdressed increased at the rate of hundreds of thousands of acres annually. In addition, lime was used, mainly in the form of ground limestone.

Another development was the conservation of surplus summer herbage as ensilage. Grass ensilage had proved itself to be not only of great value in pasture management, but also as a summer supplementary feed unsurpassed in reliability by any other summer crop.

The most significant development in grassland farming was in the strain of grass. The depletion of pasture had been due in part to the sowing of poor quality and impermanent types of grass. By contrast, when persistent leafy types of grass were sown, the modern methods of management which led to higher production were made more economical and efficient. Research on perennial ryegrass and on other grasses and clovers was well advanced. The "pedigree" grass, capable of maximum returns under hard stocking conditions (rotational grazing) was as essential as pedigree stock if grass growth was to be fully utilised. Recognition and application of this principle was destined to play an even more important role in the management of New Zealand grassland farming than any of the earlier developments.

15

TANKER COLLECTION

The introduction of technological innovations such as mechanical separation in both the creamery and farmhouse resulted in major spatial and functional restructuring in the dairy factory system. Over a period of not more than 50 or 60 years the mechanical separation process had been shifted from farm dairy to factory then to farm milkshed and back to factory floor, with comparable locational and spatial shifts in the various related processing units.

The introduction of tanker whole milk collection had an impact upon the structure, organisation and spatial patterning of the factory system, as profound as mechanical separation had been only decades earlier. During the intervening years, as the home separation system spread the dairy industry experienced steady growth, and continuing rationalisation of the production processes resulted in some factory closures, notably those proprietary concerns which had closed when milk supply areas were re-zoned to end the duplication or overlapping of neighbouring companies' collection facilities. The patterns of these earlier closures and associated changes were only minimal when compared to total reshaping of dairy factory locations, dairying landscapes, spatial distributions and relationships, caused by the impact of tanker collection.

Widespread company amalgamations which resulted in equally widespread factory closures became characteristic features of traditional dairying regions in the new tanker era. However rationalisation and specialisation meant larger capacity and multiple plant and product manufacturing complexes making the closure of smaller, less viable factories inevitable. After closure some of these factories were put to a variety of new uses, traditional as well innovative, and normally good use was made of existing sites and buildings. Dairying areas with larger numbers of small to medium sized dairy factories were most affected by closures, for example, southern Taranaki, Manawatu, Wairarapa and eastern Southland.

The full impact of tanker collection reverberated throughout the various levels of the dairy industry. Its introduction was the basis of on-farm, bulk, whole milk collection, which displaced cream can collection of the former home separation era. The newly acquired tanker fleets were also put to other uses, such as the large scale bulk transfer of milk byproducts between production related manufacturing plants. The additional functions were however, essentially complimentary; initially the whole milk was picked up from the farm and delivered to the factory,

Loading butter into a
refrigerator at Palmerston North
railway yard.

once processed, its byproducts were distributed between the various
manufacturing plants to ensure optimum utilisation of all the components of the whole milk.

Although motor vehicles had been adapted and developed for the
specialised transport of petroleum and other industrial liquids, they had
not to date proved suitable or practicable for the large-scale transport of
perishable commodities such as whole milk. The growing demand for a
means of transporting milk in bulk over longer distances, so as to permit a
larger scale and more rational utilisation of whole milk, provided the
incentive for whole milk collection. As early as 1940, in the Bakersfield
area of California, some dairy farmers had had their whole milk collected
by specially designed and equipped milk tankers.

In New Zealand, the use of milk tankers was first taken up and initiated
by the Wellington City Corporation Milk Department to transport
whole milk from its Rahui Depot, where it had been collected from the
rural area, to the city milk treatment station, where the whole milk was
prepared for consumption and distribution. But such a tanker delivery
and collection service, first introduced in 1947 was concerned with the
transfer of town milk supply rather than with tanker collection of whole
milk.

In the late 1940s, the New Zealand Co-operative Dairy Company was
transporting buttermilk by tanker between company factories in the
Waikato, while further north, Lactose Ltd was also using tankers to
transport ice-cream mix from Manurewa to Auckland; again, both
operations were concerned with the bulk transfer of milk products not
with collection.

The experience of using tankers to transfer products obviously proved
useful to the New Zealand Co-operative Dairy Company which in April

The Te Rapa tanker fleet.

1950, and with a large measure of success, introduced the first New Zealand tanker collection from farm suppliers to its Waitoa milk powder factory. The Waitoa experiment began with only 10 supply farms, all within a distance of 1½ miles of the receiving factory. Such was the early success of this experiment that by the next season the supply area had been extended to a distance of 10 miles, and by 1953, all the 550 factory supply farms were utilising the new tanker collection system.

Whole milk tanker collection was begun in 1955 in the Manawatu and in Taranaki a year later. The adoption of tanker collection progressed slowly but steadily until a period of increasingly rapid and widespread growth in the late 1950s and 1960s.

One of the principal reasons underlying the early rapid expansion of tanker collection was the growing pressure for larger scale operations and the consequent economies of scale. Even before the advent of tanker collection William Goodfellow, an entrepreneur and advocate of large scale dairy operation had brought about, as early as 1919-1920, the amalgamation of the New Zealand Dairy Association and the Waikato Co-operative Dairy Company to form the New Zealand Co-operative Dairy Company. In the following season, the incorporation of the Thames Co-operative Dairy Company created an immense new dairy

The night shift at the Morrinsville Co-operative Dairy Co.

concern, its total butter output of over 12,000 tons being more than one-third of the country's total production. Such an amalgamation was, however, scarcely comparable to the pressure for larger size, scale and capacity in both supply and production in the high cost post war era. In the 1950s and 1960s the immediate practical response, in the form of a rapid and widespread adoption of tanker collection, was more a reaction to specific external circumstances; the limitations of falling milk supply, the need to update and replace large items of plant and the need to increase levels of productivity to offset rising production costs. Thus tanker collection was essential for the future of the dairy industry. It furnished dairy companies with the means of amalgamation and the resultant benefits of scale economies, as well as scope for rationalisation, specialisation and modernisation in production.

By June 1961 58 out of a total of 132 dairy companies were making some use of tankers for whole milk collection. Their distribution throughout New Zealand was uneven but they were all closely associated with increasing the milk supply to existing or new factories as well as the transfer of milk byproducts to other plants for further processing. By the 1966-67 season 764 milk tankers were in regular operation, and of the 75% of total butterfat produced from the supply of whole milk, nearly all was accounted for by tanker.

Part of the rapid growth and expansion of dairy company tanker fleets

Automated control for
continuous butter making.

could be attributed not so much to the bulk collection of whole milk, but to the increased movement between factories of milk and milk products for manufacture. Such movements were both permanent and temporary in nature. Thus fuller utilisation of byproducts in milk processing was more readily carried out and facilitated by tanker transfer while short-term shifts of whole milk supply could offset temporary hold-ups or imbalances in processing.

Tankers simplified the whole process of collection, and delivery of milk supply but at the same time created an immediate need for specialised facilities. The farmer had to provide on-farm access roads from farm gate to milkshed, cattle stops, and covered stands for the holding of storage vats. In some cases existing milkshed plant had to be modified for milk delivery into the vats. For example, collection times were now critical while tankers and vats rapidly became highly specialised plant that tended increasingly to dominate farm operations.

Factory layout and design of plant had to be significantly modified. The former factory receiving stage, where cream cans had been received, weighed, emptied and washed before being returned now became obsolete. In its place were special tanker receiving bays, designed to expedite the bulk receipt of whole milk and the rapid turnaround of tankers and trailer. A single tanker with a 1,500-1,800 gallon capacity per load replaced the flat top truck with its 1,200 gallons in large numbers of small cans. One company claimed that the use of tankers instead of flat top trucks, had enabled it to pick up 125,000 gallons more milk while at the same time its drivers had travelled 11,000 miles less in distance. Tanker collection therefore resulted in major transport economies for farm and factory although immediate costs actually increased with the purchase and installation of specialised tankers and trailers, vats and

Continuous butter-maker.

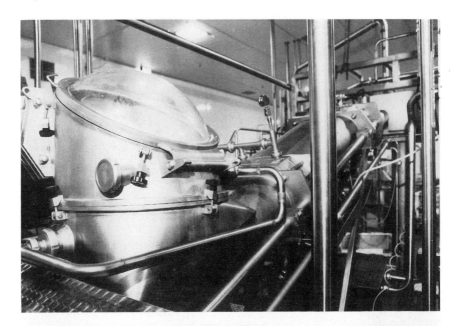

Packing hoppers.

tanker bays. Overall, though, costs were kept at a relatively low level and vehicle maintenance and installation more than offset the physical handling, returning and frequent re-tinning of thousands of individual cream cans.

There were, of course, some disadvantages. One supplier's milk could easily contaminate an entire tanker load, while the responsibility for decisions on the acceptance or refusal of such milk was shifted from the qualified factory grader to the tanker driver, already under the pressure of maintaining a tight collection schedule. This situation was eased by the introduction of refrigerated holding vats for cooling and storing milk on the farm, and by the later introduction of "skip-a-day" collection, which gained transport economies by collecting milk only every second day

from more isolated suppliers with adequate milk storage and cooling facilities.

A further difficulty related to the measurement of milk supply from the individual farms which at times resulted in great dissatisfaction among factory suppliers. With cream can collection, weighing on the receiving stage had provided a precise measure, but tanker collection involved only a visual calculation by the driver on a graduated indicator tube on the tanker.

Tanker collection resulted in major changes in factory operation and company organisation. The scale of factory operations, measured by volume and area of milk supply, as well as the quantity output, increased markedly. The quantity of butterfat supplied to and processed by New Zealand dairy factories rose from 4.2 million lbs in 1951 to 8.8 million lbs by 1969. Butter, cheese and milk powder output increased correspondingly. There was also a much more efficient utilisation of the total milk supply, as well as whole milk content, such as milk solids, in dairy products other than the traditional butter and cheese.

Company organisation and management was also influenced for amalgamations functionally linked many widely scattered production units. Before this era of widespread tanker collection, the long distance transport of whole milk or milk byproducts had seemed uneconomic and impracticable. Consequently, dairy companies had generally remained relatively small scale operations, the New Zealand Co-operative Dairy Company being the only notable exception. Companies also tended to be concerned with a single product and dual plant factories were rare. With

Milk in a 2600 gallon vat being cooked and agitated during cheese making.

149

Modern separators.

tanker collection greater and more extensive inter-factory contact was not only inevitable but encouraged and facilitated. This resulted in the widespread recognition within the industry that company amalgamations were practicable and extremely viable alternative propositions for future development. Such amalgamations significantly increased production capacity.

Tanker collection enabled the development of two quite distinct spatial forms of large-scale dairy factory organisation: the "Waikato" type of polycentric and multiple branch factory pattern, and the "Manawatu" or "Whareroa" type of a single site with multiple plant and product. Thus, in the Waikato the New Zealand Co-operative Dairy Company operated a widespread series of separate, highly specialised but functionally linked factories under the central administration and control of a single company. Tanker collection and inter-factory movement facilitated the collection of liquid milk or milk byproducts at various processing or production centres for further manufacturing.

By comparison the Manawatu model, involved negligible inter-factory movement of raw material. This was because dairy plant and production were already concentrated upon a single site that incorporated multiple plant and production processes. Thus Manawatu in its Longburn complex produced butter, milk powder and casein in its major plants, together with lactose.

This ability to move whole milk or byproduct supply between factories or divert supply to alternative plants was significant in increasing flexibility in operations because it is vital to maintain a pattern of regular and uninterrupted production flow, even at the expense of less economic tanker movements. The cost of shutting down factory operations or the

loss of partly processed products was much greater than an under-utilised tanker fleet.

Old and new bottle filling machines.

The development of much larger dairy companies, with comparably expanded scales of operation, resulted in other changes, particularly in the relationship between factory suppliers and factory management. Although the co-operative organisation continued, the relations became much more formal as the larger company pattern displaced the smaller company organisation and the close supplier-manager association. Work became more specific and specialised, tasks more sophisticated and technical. Even tanker routing became a very precise and demanding routine mathematical exercise. More specialised and highly qualified staff were employed in new company operations — vehicle drivers, maintenance mechanics, engineers, plant technicians, market analysts, laboratory assistants, computer operators, to mention but a few. Automation and computerised controls became a feature of factory and plant operations requiring skilled and qualified operating staff. Policy and decision-making lay increasingly with directorates meeting upon the advice and recommendation of specialist personnel. Factory management became less involved in such decisions with the result that suppliers had less direct influence, although the ward system of representation helped to redress this situation, giving the supplier greater access and representation. With the increasingly capital intensive nature of the plant, factory and company operations, shareholders increased in number in response to growing capital needs.

Change also occured at the farm level. With a marked decline in the availability of skim and other milk byproducts for pig and calf feeding, pig raising disappeared on many dairy farms, making them less flexible and

The milk powder control room at Te Rapa.

more specialised. Dairy herd sizes tended to grow as total herd numbers declined, and there was a continuing pressure for greater capital investment to increase productivity levels, for example, the use of rotary milking parlours. Overall, dairying became increasingly specialised and more capital intensive.

The requirements for milk supply were different for town milk, butter, cheese, casein and milk powder production, and accordingly the relative advantages of tanker collection varied significantly between these major branches of dairy production. The general advantage gained from whole milk tanker collection was its specific impact upon the radius and size of each factory's individual collection and supply area, and the economy of time and labour. Most factories had their potential milk supply areas significantly extended and similarly their volume of milk supply expanded.

Cheese making factories needed a regular and reliable supply of whole milk every day. Before the introduction of tanker collection the volume of milk available was limited. This restricted the production capacity of the typical cheese factory. As a result there was a relatively large number of small to medium sized factories, each producing as little as 200-300 tons of cheese annually. The factories were evenly spread and were often located as little as 5 miles apart.

Tanker collection extended both the possible collection area and the volume of milk supply for all such cheese factories. Thus in the predominantly cheese producing areas of southern and central Taranaki, and the Mataura Valley in eastern Southland, a new spatial pattern

Milk powder packing room at Te Rapa.

emerged. There were far fewer factories but the capacity of those that remained was much larger; they were spaced at less regular and longer intervals along the main roads in the cheese producing districts.

Tanker collection demonstrated the practicalities and possibilities of increased milk powder production in those dairying areas where the previous milk supply had proved too small and limited and farm settlement too sparse and inadequate to sustain a factory. To operate a sizeable milk powder plant large quantities of whole milk or alternative raw material were needed on a regular basis. By extending the supply area tanker collection assured just such a supply. Some milk powder and casein plants developed in close association with butter or cheese plants.

Town milk suppliers, once given the option, turned quickly to tanker collection because it met the stringent demands on quantity and quality. There was, therefore, a rapid changeover to tanker whole milk collection by cheese, casein and milk powder factories and town milk stations. In contrast, the creameries were much slower to adopt tanker collection. The home separation system had already provided the creameries with the very real advantage of requiring a much smaller volume of raw material. The amount of cream required was about one tenth of that needed had they been processing milk. Another factor which was very important was their supply network. A significant amount of the cream delivered for processing came from smaller, casual or part-time suppliers who would have found it uneconomic to install the plant necessary for tanker collection. The use of tanker collection was espoused more quickly where cream for butter manufacture was only a part of the total operation. Where billy can and whole milk tanker collection were promoted together, it was customary to have two collection systems, with tankers often equipped accordingly, although truck collection of cream cans by company or contractor tended to persist in areas of predominantly cream supply. The spread and extension of tanker collection was provided for by the introduction of smaller capacity tanker trailers in association with the tanker unit, areas of whole milk supply tended to be serviced by tankers while billy cans remained only for more isolated creamery supply.

CHAPTER

16

COMMANDEER

AND COMMON MARKET

In 1881, on the threshold of a major advance and development in factory dairying, the total value of butter and cheese exports was £14,600 or only 0.25% of export values. Despite the apparent breakthrough of 1882, the embryo industry responded only slowly to the real possibilities of an export dairy trade. In the 5 year period from 1882-87, when total export values rose from £6.5 million to nearly £7.5 million, the total value of both butter and cheese exports rose from only £62,200 in 1882 to £197,200 in 1887. The 1880s were, however, years of agricultural recession when even edible fungus exceeded dairy produce in export values.

Over the next 40 years the value of cheese and butter exports together with their overall percentage of total export values steadily increased as factory production expanded and export facilities grew. The dairy export trade grew from a value of about £350,000 in 1893 to a huge £17.4 million in 1933.

Early dairy sales, and the dairy export trade generally, were conducted by individual dairy companies and proprietors through well-established agents whose offices and coolstore facilities were located in Tooley Street, London's traditional butter market on the south bank of the River Thames. The dairy produce was all sold on the free marketing system with the directorates or proprietors of individual companies accepting full responsibility for sales and trading. They decided upon terms of sale, methods of transport and choice of agents.

Tooley Street was a term which embraced the many agents, merchants and brokers engaged in the dairy produce import trade in Britain. All the leading dairy traders and major outlets in Britain were represented there. Tooley Street was lined with warehouses, many of them fitted with coolstores and abutting onto the wharves along the Thames. It included a total of 27 firms handling dairy produce from New Zealand, 26 of which were members of the New Zealand Dairy Produce Importers' Association, acting for importers generally, and the other, Empire Dairies, formed by William Goodfellow in 1928. In addition there were several wholesale firms handling small quantities of butter from individual New Zealand dairy companies on a consignment basis. Each of the Tooley Street merchant firms had representatives in New Zealand to keep in contact with those companies whose produce they handled and also to promote new dairy business.

On the Thames waterfront, adjacent to Tooley Street, were lines of wharves, Hay's Wharf, established in 1770, was the principal Tooley Street dairy produce wharf. It was equipped with a fleet of 350 river tugs and lighters to link up with ocean going steamers discharging their cargoes of dairy produce at the Royal Albert and other docks situated some miles downriver from Tooley Street. Under a 1928 contract with the New Zealand Dairy Produce Board, the proprietors of the privately owned Hay's Wharf were responsible for the unloading, transporting, storage and distribution of all the trade in dairy produce shipped to the port of London.

The free marketing system had serious deficiencies and in 1914, before the outbreak of war, plans were underway for an industry wide collective marketing scheme. However, the special wartime circumstances and the need to ensure the proper channelling of all available surplus food supply to avoid unnecessary shortages in Britain, resulted in the British Government imposing a commandeer on all New Zealand dairy produce. This was the first time ever that Government to Government trading in dairy produce had ever taken place between the two countries. During the first 2 years of war the requisition order applied to only a part of New Zealand's annual cheese output but from 1916-18 the food situation in Britain became so desperate that it necessitated the extension of the commandeer order to all surplus butter and cheese. Such an arrangement continued until 1921 when all bulk purchase ageements were terminated and the free marketing system was re-introduced.

The rapid growth in the numbers of small cheese factories was due mainly to the price differential favouring cheese and to the marketing policy that clearly identified cheese as the ideal wartime food import. For 7 years the commandeer provided the stability of predetermined dairy produce prices and an assured market, but in 1922 fierce competition began once more as New Zealand, along with other major dairy exporting countries, joined battle for rights of access to the British market. As virtually the only world dairy market still open to exporting countries it was of particular importance.

Oversupply, among other factors, had resulted in drastic falls in prices in the early 1920s and the New Zealand dairy industry leaders made further attempts to organise a collective marketing system that would ensure higher returns to producers. The first scheme involved the formation of a joint New Zealand and English limited company to pool all produce and market it in Britain. The company began to operate in 1920-21 with 70 member dairy companies and within 5 years membership had risen to 91. Even as this scheme was being implemented, the New Zealand Co-operative Dairy Company, already an extremely large New Zealand exporter, was operating its own marketing organisation in London, allocating the company's produce to Tooley Street firms or to agents to sell under supervision.

In 1922 the dairy industry in New Zealand set up the Dairy Council to promote industry legislation for the central marketing of all dairy produce. An Act of Parliament passed in 1923 established the New Zealand Dairy Produce Export Control Board (the forerunner of the New Zealand Dairy Board) which had extensive powers for the regulation of marketing. Its policy "absolute and total control" was introduced in 1926 and based on the premise that dairy produce from New Zealand should be sold only through selected agents in Tooley Street. A minimum price tag was to be determined by the Board.

New Zealand producers thought the policy fair and reasonable but it aroused intense opposition from British business and marketing interests on the grounds that it infringed free trade. So effective were the latter

New Zealand dairy produce on display at Liverpool.

groups in mustering opposition that the price fixation policy was abandoned in 1927 after only a single marketing season. However, the basic principles for promoting an orderly marketing system were retained as was the notion of keeping a greater share of the returns for the producers. These basic principles included: regulating and timing shipments to ensure a more uniform and regular market supply; recording all dairy sales, agents' names and prices paid for produce; arranging freight and insurance contracts; and introducing a nationwide

The Tooley Street headquarters of Amalgamated Dairies.

dairy produce brand, Fernleaf, and a finest grade of butter and cheese with a price premium for each.

In 1927 the New Zealand Co-operative Dairy Company began to reorganise its own marketing organisation, inviting other dairy co-operatives to join with it in a marketing co-operative that aimed to achieve, voluntarily, what the Board scheme had failed to accomplish by compulsion. At the end of 1927, the new company, Amalgamated Dairies Ltd, began selling through Tooley Street agents under the supervision of the company's London manager, but without policies of price fixing or the withholding of market supplies.

In 1928 Amalgamated Dairies Ltd linked up with the Australian producers' organisation to form Empire Dairies Ltd, the ultimate aim was to market all Australasian dairy produce in Britain. Empire Dairies not only continued the former policy of allocating dairy produce to Tooley Street merchants, but also set itself up as a wholesale agency in opposition to Tooley Street firms, aiming to obtain for its producers a share of the profits derived from Tooley Street sales to commercial distributors. Amalgamated Dairies ceased operations but Empire Dairies continued to function as a wholesale agency and until the introduction of the guaranteed price scheme, handled the marketing of all the New Zealand Co-operative Dairy Company's produce.

The failure of the Dairy Produce Export Control Board's export policy of the 1920s discouraged any further efforts at national marketing schemes until the depression years of the 1930s when, before a new marketing scheme could be implemented, there was a change of Government. In 1935 the Labour Government took office and immediately proceeded to implement plans for a guaranteed price scheme. This was introduced in 1936 and the principal provisions were: the Government was to acquire all butter and cheese from companies at a predetermined price fixed for the season and based on an estimated cost of production; and the Government was to sell through approved agents in Britain, paying a 2% commission on prices realised — there was no attempt at even minimum price fixation.

In the first year of the guaranteed price scheme the Government based

Cheese for export loaded onto
the ship from insulated railway
wagons.

the prices to be paid on the previous 10 years. This was criticised by the industry since the period included rather low prices — a result of heavy production to offset the effects of the depression years. From 1937-38 the guaranteed price was based upon production costs, largely exclusive of realisations, but bearing in mind the need to maintain stability and efficiency through effective pricing. Finally the basis of the calculation was agreed upon as the operating costs of a standard farm, employing two labour units and running 48 cows with an average production of 250 lbs of butterfat per cow. This basis, much resented by the industry, remained in place until 1956 with only annual adjustments to keep it in step with internal cost movements.

In 1939 the British Government once again adopted special wartime control measures to ensure a continuing food supply. The marketing system changed completely. The British Ministry of Food became the sole purchaser of imported dairy produce and entered into direct negotiations with the New Zealand Government for the supply of all surplus dairy produce. This was disposed of through existing marketing agents who received a commission for their services. Initially the scheme was based upon annually negotiated contracts between the two Governments, with the purchase price determined by the costs of production in New Zealand. Although nominally covering all exportable surpluses each year it was agreed to leave a percentage of the so called surplus to sell on other markets to ensure they would be retained for future exports. In 1944 the annual basis of negotiation was replaced by a 7 year contract.

Long before the 7 year contract expired, the Government yielded to an industry request to remove marketing negotiation from direct Govern-

Glaxo Laboratories, famous for milk powder.

ment control to an independent body representing both the Government and the industry itself. In 1947 the New Zealand Dairy Products Marketing Commission was established with functions identical to those of the Government's Marketing Department, and although it was nominally a non-political body the balance of power lay with Government appointees. Its functions included acquiring, marketing and promoting dairy produce intended for export, fixing the price to be paid, and controlling the marketing of butter and certain kinds of cheeses within New Zealand.

The 1944 agreement was re-negotiated in 1948 for a period of 7 years, one of the terms being that the price to be paid was not to vary more than plus or minus 7.5% of the previous year's price. Contracts were also entered into for the supply of milk powders and other dairy products. With post war costs spiralling in New Zealand, the full 7.5% increase was agreed to in every year but one. In 1954 New Zealand and Britain mutually agreed to terminate the contract — New Zealand believed it would result in better prices and Britain wished to do away with controls over food supply feeling that domestic production together with imports could assure adequate provisions for its population. With the British Government no longer involved in dairy trading after 1954, the marketing system reverted to that of pre war years with a Marketing Commission replacing the New Zealand Government's Marketing Department.

In 1960 joint meetings between the Dairy Produce Marketing Commission and the New Zealand Dairy Board were held to discuss means of avoiding divided control in the industry. As a result the Dairy Production and Marketing Board was set up in 1961, its name being abbreviated in 1966 to the New Zealand Dairy Board.

Marketing was based traditionally on butter and cheese and the historical basis of the guaranteed price system was the price a farmer received for each pound of butterfat used in butter manufacture. The initial cheese differential was intended to compensate cheese factory suppliers for the loss of skim milk normally retained for use in pig and calf raising and for additional milk cartage costs. More recently the differential has compensated cheese factory suppliers for the loss of income from valuable byproducts, mainly casein and skim milk powders, now

Blue vein cheese in the dry curing room.

essentially complimentary income sources.

Until 1954 bulk purchasing was a feature of the New Zealand-United Kingdom dairy trade but its end was not unwelcome. Locally it was a time not only of rising production and productivity but also one of rising prices. There were fresh inputs of capital and technological expertise into dairying. Although cheese exports in the period 1954-58 averaging 95,000 tons annually, were only 3,000 tons more than for the period 1934-38, butter production had responded to innovations in quality and packaging so that average annual exports of 162,000 tons for the years 1934-38 had risen to 207,000 tons.

By 1958 however, butter consumption was coming under threat from cheaper vegetable oils and margarine as well as a growing awareness of the effects of cholesterol. New Zealand, with the highest per capita butter consumption in the world, was increasingly concerned. In the USA Americans halved their butter consumption from pre war years and production fell by 30% with a corresponding increase in margarine.

The British dairy market in the 1950s was the only accessible dairy market of the time, absorbing as much as 425,000 tons or over 80% of the butter entering the international dairy trade. In such a situation competition for a share of this market from overseas as well as domestic suppliers who were expanding production rapidly, was intense. Late in the 1950s a potentially huge dairy producer emerged as a result of the Treaty of Rome which saw the formation of the Common Market. This development caused grave concern to those dairy industries dependent upon international dairy produce.

No-one questioned the need to reorganise Europe's traditional farming

Dried milk powder packed for export.

practices, the concern arose from the manner in which it was proposed to protect and rationalise Europe's dairy industry and the implications for producers beyond the tariff wall. The desirable target prices were set, initially only nationally, but later as part of a common system and based on the costs of existing producers. In addition there were official minimum, or intervention, price levels below which support was provided. Official funds were used to buy storable dairy surpluses at agreed prices and to subsidise or refund dairy exports as necessary. Imports of dairy produce were actively discouraged by means of levies and quotas upon external suppliers. Such imports as were permitted could be sold only at the basic target price and funds accruing in this manner were used to subsidise and assist European dairy producers. Whatever the original intention of such policies, the effect was to stimulate local production while pricing out increased local consumption resulting in further dairy surpluses — the so called butter mountains. In the period between the termination of New Zealand's contract with Britain in 1954 and the formation of the EEC in 1957, Europe's butter output had risen from 832,000 to 930,000 tons and protectionist policies would cause this to increase even more.

So ended the era of free market and voluntary restraint in dairy exporting. When Britain finally became a member of the Common Market New Zealand was offered quotas on the British market and became a "special case" as the common agricultural policy developed. West Germany, formerly the world's second largest dairy market ceased to be so and pressures on the British market were exerted from both France and the Netherlands. The British butter market was unable to absorb the increasing quantities available and the cheese market too was seriously affected.

The need for New Zealand to find new markets became essential although efforts in this direction were largely frustrated by EEC and USA protectionist policies as they tried to come to grips with their chronic dairy surpluses. At the same time, butterfat processing became significantly more varied and refined. Automatic continuous processes began to replace the more traditional batch processes. Butter churning was displaced by continuous processes, while packaging became more and

At the docks awaiting loading of dairy produce for export.

more consumer-conscious. New Zealand's basic salted butter was supplemented by unsalted and ready-spread varieties as well as some adapted to tropical heat. Alternative milk-fat uses and outlets were also developed — ice-cream mix, frozen cream, anhydrous milk-fat (AMF) and ghee. A double shift from butter to butter-oil and British to Third World marketing entailed a significant revenue reduction.

Cheese, the other major butterfat product followed trends similar to those of butter. A protein-rich product, cheese had been more independent of the London market, but its marked cheddar bias was a notable handicap in taking advantage of increased cheese comsumption. A few early cheese-makers had tried labour-intensive swiss and stilton, but soft cheeses tended to spoil in transit while standard, mass-produced cheddars commanded the British market.

In the 1960s as British and Irish production expanded, the Europeans turned more and more to cheddar production and exports, and the New Zealand industry found re-tooling essential to meet the changing market demand. The American trade needed colby and monterey; the Japanese wanted gouda. The factories initially built and equipped for cheddar were also able to manufacture colby, cheshire, granular and skim-milk varieties — but brine-salted cheese such as gouda and Italian varieties needed new plant. Edam, leidanz, danbo, munster and gruyere were added, while, with an eye to a future Japanese market, cheddar and gouda were blended into a distinctive "Egmont cheese". Cheddar was also emulsified into a processed cheese.

Such diversification was, nevertheless, fairly limited so that by 1972-73 cheddar still accounted for 86% of the export tonnage, "other kinds" some 13%, and "processed" only 1%. Market diversification however, gained momentum. Over the period 1969-73, the proportion of New Zealand cheese production exported to the British market fell from 80 to 63%, while at the same time the proportion of butter fell from 96 to 86%.

Butterfat was complimented in whole-milk by a subtle complex of non-fat milk solids. Post-war dairy processing technology increasingly permitted more of these various components to be identified and analysed, and subdivided and reconstituted. New Zealand's dairy industry set out to supply both milk solids and processing technology in Southeast Asia. Tinned condensed milk was popular but the new industry blended New Zealand milk powder components with tropical cane sugar. National governments and multi-national corporations showed real interest and in 1961 New Zealand's initial plant opened in Singapore, followed by others in Mauritius, Southeast Asia and the West Indies. Plants and retail outlets were soon transferred to local ownership, however, although New Zealand sought to retain a guaranteed role as raw material supplier and technical adviser. Hence an increase in exports of both milk-fat and non-fat solids and growing trade with Mexico, Peru and Chile. Dried butter-milk and especially skim-milk powder were the key components in this expansion and were also naturally complimentary to the existing massive butter industry. The production of 10 lbs of butter also yielded as by-products about 100 lbs of skim-milk by cream separation and a pound of butter-milk by churning. The butter-milk could be dried into a powder, while skim-milk produced a powder or casein. The potential for diversification into non-fat milk solids production increased as dairy factories and companies acquired ever-larger quantities of fresh whole-milk as well as the control of its handling and processing with ever-greater levels of technical sophistication. Initially milk-drying was carried out by the economical roller-drying process, the powder process was more refined and costly, although necessary for human consumption. Dried butter-milk, and more especially skim-milk powder production, expanded rapidly as spray-drying outpaced roller-drying, especially for skim-milk processing, as the balance shifted from animal to human nutrition and trade and market shifted from British to Asian and Latin American outlets.

Roller-dried butter-milk powder was prepared largely for livestock feeds, while spray-dried butter-milk was an excellent smoothing ingredient in reconstituted milk. Overall production growth was modest, but whereas in 1962 some 44% was roller-dried, with 53% destined for Britain, by 1972 some 79% for export was spray-dried, with only 12% roller-dried for the London market.

Skim-milk powder was not only much more abundant but equally more versatile. It could add finish to livestock feed, nutrition to confection and beverages, plumpness to sausages and crustiness to bread. While "instantising" accentuated its appeal in western markets, its main use for Third World consumers was milk reconstitution. New Zealand roller-dried skim-milk powder was mainly for home and British consumption, but the spray-dried product was soon dominant with the Asian market being crucial. Tailoring for the specification trade became a New Zealand speciality, a positive response to export opportunities. Milk powder sales to Britain slumped badly by the 1970s, but total exports,

Modern cheddar master cheese-maker.

almost all spray-dried, surged from 43,400 tonnes in 1962 to 158,300 in 1972 with major markets in Japan, the Persian Gulf, Latin America and the Caribbean.

Skim-milk powder, as one component in the milk recombination complex, followed much the same trading channels as dried milkfat and butter-milk, but casein, the alternative skim-milk product, followed a separate course, usually to the more industrialised markets. Here it filled various different niches — coagulated by renin extract it was mainly used for costume jewellery, but precipitated by lactic acid it yielded a versatile product utilised in glues, fibres, foodstuffs and especially printing paper.

In addition to the diversification programme, in the early 1970s New Zealand launched a diplomatic campaign, or rather offensive, without precedent to try to foster a favourable climate of opinion within the EEC Diversification, true enough, had made great progress, but the real threat was still the possibility of total exclusion from a still significant market, for in 1970 the British dairy market still accounted for 53% by volume and 57% by value of New Zealand's dairy trade. That market was doubly significant — Britain remained the largest single buyer overall and the only consistent bulk buyer of butter and cheese. These were pivotal elements in the dairy trade, both in themselves and also in their relationship to skim-milk, butter-milk, casein and whey, so essential for any diversification.

INDEX